CHEMOMETRICS
Applications of Mathematics
and Statistics to Laboratory Systems

CHEMOMETRICS
Applications of Mathematics and Statistics to Laboratory Systems

RICHARD G. BRERETON B.A., M.A., Ph.D.
Department of Chemistry, University of Bristol

ELLIS HORWOOD
NEW YORK LONDON TORONTO SYDNEY TOKYO SINGAPORE

First published in 1990 by
ELLIS HORWOOD LIMITED
Market Cross House, Cooper Street,
Chichester, West Sussex, PO19 1EB, England

A division of
Simon & Schuster International Group
A Paramount Communications Company

Printed and bound in Great Britain
by Bookcraft (Bath) Limited, Midsomer Norton, Avon

British Library Cataloguing in Publication Data

Brereton, Richard G.
Chemometrics: applications of mathematics and statistics to laboratory systems.
1. Chemistry. Applications of statistical mathematics. 2. Laboratory techniques.
Applications of computer systems.
I. Title
540.15195
ISBN 0–13–131350–9

Library of Congress Cataloging-in-Publication Data

Brereton, Richard G.
Chemometrics: applications of mathematics and statistics to laboratory systems /
Richard G. Brereton.
p. cm. — (Ellis Horwood series in chemical computation, statistics, and information)
Includes bibliographical references and index.
ISBN 0–13–131350–9
1. Chemistry, Analytic — Statistical methods. 2. Chemistry, Analytic — Mathematics.
I. Title. II. Series.
QD75.4.S8B74   1990
543'.0072–dc20
90–43929
CIP

Table of contents

Preface

There has been an enormous upsurge in interest in chemometrics over the past few years. This has been catalysed by several events which include the founding of two international journals, several important meetings, the development of courses, and the creation of various national groups such as the UK chemometrics discussion group.

Originally regarded as a rather specialized branch of statistical analytical chemistry, chemometrics has broadened in scope and interest. User-friendly software has brought chemometrics methods to the non-expert, and so there is an urgent need for the laboratory based scientist to understand the use of and need for chemometrics. Together with the increased availability of chemometric software has come the increased realization that chemometrics has far broader applications than analytical chemistry: a synthetic chemist wanting to optimize the yield of his reaction, a spectroscopist trying to improve the quality of a spectrum, a kineticist trying to working out complex reaction pathways and a pharmacologist trying to understand the basis of activity of a drug all need to use chemometric methods. Common to all these applications is the ability of the laboratory based experimenter to acquire huge quantities of data using modern instrumental techniques. Indeed, the applications of chemometrics spread wider even than chemistry and are useful to all laboratory based research workers.

In a subject like chemometrics there will, inevitably, be a gap between the specialist developer of chemometric methods, often trained in statistics and computing, and the user of these methods, who does not necessarily have the time or background to understand the mathematical basis of the subject but who needs to know what is possible and what he or she can safely do before consulting an expert. This book is unashamedly aimed at the latter user of chemometrics and does not attempt to provide comprehensive information about computational algorithms or mathematical statistics. A feature of this text is extensive, simple, numerical examples. Most of these have been produced using a spreadsheet (Wingz on an Apple Macintosh) without recourse to any specialized software. Rather than the examples being presented in the form of source listings or spreadsheet macros they are in the form of tables, enabling the reader to reproduce the data and explore the main trends using his or her own favourite software.

Chemometrics is still a very controversial subject, and the relationship between statistician and experimentalist is in its early stages of development. I have metamorphosed over the last decade from an experimentalist whose Ph.D. was dependent on extraction, separation and structure determination of unstable compounds to a computer based chemometrician. This experience has taught me the importance of integrating experimentation with analysis. The chemometrician should understand the difficulties of hands-on laboratory experimentation as well as the statistical basis of chemometric techniques. Over this last decade there have also been dramatic changes in scientific computing. The user of 1980 was likely to be a mathematically minded programmer running large calculations on off-line mainframes; the scientific computer user of 1990 is more likely to be running fast and user-friendly software on a benchtop microcomputer. Those of us who have lasted in scientific computing over the last decade have had to change with the times to survive. Instead of sitting in offices devising new algorithms in FORTRAN, we have had to learn to go out and sell our methods to the potential user. Many of the most prominent chemometricians invest much time in running courses, writing articles, organizing meetings, writing books, developing user-friendly software and so on, although there will always be the need to develop new computational methods.

The approach adopted in this text is the product of several years' scientific experiences, both pleasant and unpleasant. On the bad side I will never forget the passion with which some established scientists reject chemometric ideas. Many scientists invest expensive resources for the collection of data and do not like being told, several years after they have purchased costly instrumentation and spent many manyears on a set of experiments, that their experiments are poorly designed, or that their instrumental method of analysis is not sensitive enough to provide the quality of conclusions they want. These experimentalists will turn to whoever is prepared to confirm their preconceptions. On the positive side, I acknowledge with pleasure the enthusiasm and collaboration of my colleagues in the UK chemometrics discussion group: within 5 years we have worked towards a general acceptance of chemometrics in the UK with some success. It has also been a pleasure to associate with the development of the journal *Chemometrics and Intelligent Laboratory Systems*. Finally the Bristol Spring School in chemometrics has been a great inspiration to me, both from interacting with my fellow tutors and from the regular and critical feedback from our students; collaboration with Bristol's Department for Continuing Education, *via* surveys, joint projects, planning courses, and even market research, has also helped me work out what people want from the chemometricians.

I thank innumerable people who have read various drafts of various chapters, particularly J.C.Berridge, H.Smit and O.M.Kvalheim. I thank Elsevier Science Publishers for copyright permission to reproduce the following figures: 2.23, 2.24, 2.26, 2.27, 2.28, 2.29, 3.9, 3.10, 4.18, 4.19, 4.21, 5.6, 5.7, 5.8 and 7.24, and *Nature* for permission to reproduce Fig. 5.21. Finally I thank the publishers for patience during the production of this book.

Richard G. Brereton
August 1990

1

Introduction

1.1 HISTORY AND PHILOSOPHY OF CHEMOMETRICS

The first people to call themselves chemometricians were chemists by training. The word chemometrics was first coined by the Swedish physical organic chemist, S. Wold, when submitting a grant application in 1971. Soon afterwards he collaborated with the American analytical chemist, B.R.Kowalski, who had been working on methods for pattern recognition in chemistry, to found the International Chemometrics Society in 1974. Although these events are officially regarded by some as the starting points of the discipline of chemometrics, the application of statistical and mathematical methods as an aid to the interpretation and acquisition of chemical data has had a long and diverse history, and it is only in the late 1980s that several diverse strands have merged under one umbrella. Other aspects of chemometrics, such as development of clustering methods in clinical analytical chemistry, use of simplex optimization to improve instrumental performance, and development of fast filters in spectroscopy, have been reported throughout the chemical literature in the 1960s and 1970s. These strands are equally part of modern chemometrics.

The statistician views chemometrics somewhat differently to the chemist. There are many well established branches of applied statistics, such as psychometrics, biometry and econometrics, and chemometrics is a relatively recent and, as yet, less well developed, application of statistics. Few psychologists, economists and so on are schooled in numerate science and so many need to consult statisticians so as to design and analyse experiments. This process of consultation has played an important role in the development of, for example, psychometrics. Experimental data are normally expensive to acquire, so statistical data analysis is absolutely essential to interpret the results of psychometric tests. A symbiosis has developed between the applied statistician and the experimental psychologist, and special statistical approaches have been developed to cope with the unique problems of the psychologist. Similar comments can be made about the development of statistical methods in sociology, economics and so on.

A major difficulty with soft sciences such as psychology is that there are severe difficulties with sampling, and often limited possibilities of reporting measurements. As we progress towards the harder scientific disciplines, we find that although biometrics, medical statistics and mathematical geology also have a long and distinguished history, because of the greater ease with which models can be developed and tested, different statistical and mathematical approaches have been emphasized. Many techniques developed for clinicians and geologists have a direct relevance to chemometrics and many textbooks written for scientists in these fields provide excellent introductions to applied statistics for experimental scientists. Within the area of laboratory rather than field based subjects, there has been much less collaboration with applied statisticians. General statistics is rarely taught as part of chemistry courses and a knowledge of such methods has not previously been thought essential to the professional chemist. Some chemists, such as food chemists, geochemists, pharmacologists and environmental chemists, have picked up statistical methods as they venture into applications areas and have tried to apply these approaches to chemical data, but on the whole the chemical community remains resistant to collaborating with statisticians.

One of the problems is that chemistry is a mixture of quantitative and descriptive science. Within areas such as quantum mechanics and statistical thermodynamics, mathematical and computational methods are well established. On the other hand, most organic chemists choose to work in this area of chemistry principally because they are not interested in mathematical methods. So the chemistry community has remained rather insular. If an organic chemist wants to perform a complex calculation, he will probably consult a theoretical chemist in the next corridor or else go down to his departmental library, rather than make an appointment with, telephone, or visit, a statistician. If he or his in-house colleague has not heard of a particular applied statistical method, he is unlikely to have the time to surmount the departmental barrier and consult an applied statistician, who may well not know what a ketone or a chromatogram is. This contrasts with the biologist, who will almost certainly have to look outside his department for advice on numerical methods, and who will therefore cultivate good relationships with the statistician.

So there has developed a communication barrier between the statistician and chemist, which is hard to break down. Yet simultaneously there has developed an enormous need for statistical methods in chemistry, for two reasons. The first is the ability of modern computerized instrumentation to acquire huge quantities of data. What is to be done with the data? Are they to be stored in filing cabinets? Are they to be deleted from disc? People would like to be able to interpret these data better, and so instrument manufacturers are slowly incorporating chemometric software into computerized instruments. The second reason is the desire to study the chemistry of mixtures. Many processes involve extremely complex mixtures: these include pharmacological, industrial, geochemical, environmental and agricultural processes. What is the proportion of different compounds in these mixtures? What are the main reactions taking place? Traditional approaches of isolating each single compound, elucidating the structure, and then developing a method for estimating the concentration of the compound in each sample are clearly extremely time consuming and not really practicable in many situations. Computational analysis of the spectra and chromatograms of mixtures helps solve these problems, and helps the chemist model complex processes.

At first it might be thought that all that needs to be done is to convince the chemist to open up books and use software developed for the applied statistician and all will be well. Unfortunately laboratory based data pose different problems to data in most other areas of applied statistics such as econometrics. There are three principal reasons for this.

(1) *Multivariate nature of chemical data*. In traditional areas of statistics, measurements are normally expensive relative to experiments. For example, we might want to examine the average size of foxes in relation to sex, season of the year, age, environment and so on. The way to do this is to physically measure the dimensions of the fox. There are not very many measurements we can make, particularly on live foxes, and each measurement takes time. In contrast, in chemistry it is possible to record several hundred pieces of information such as spectroscopic and chromatographic intensities on each sample. Therefore, most chemical experiments are *multivariate*, that is several pieces of information can be obtained per experiment. Indeed, in many cases, more variables (measurements) are available than samples (experiments). Much traditional statistical software will not cope, or is not designed to cope, with such wealth of information. New chemometric approaches need to be developed.

(2) *Experimental error*. This will be a recurring theme of this text, and is fundamental to the type of statistics developed to tackle a given problem. In traditional statistics, sampling errors are normally relatively large, and reproducibility of experiments corresponding low. Consider, for example, using psychometric tests to link verbal ability of school children to age. Although there will be some correlation with age many other factors will influence the performance of school children, and, so, if we plotted a graph of scores on a verbal ability test against age of pupil, we expect high scatter, and low reproducibility. In contrast, it is normally possible to obtain

relatively high reproducibility of replicate analysis of chromatographic intensities. But, chemometrics should be contrasted with traditional physical chemistry, in which experiments are always better than theory. In quantum mechanics, because the sample size is very large (typically of the order of 10^{23}) experiments are highly reproducible, and with good apparatus can measure such parameters as energy of bonding of two atoms with extremely high accuracy. In chemometrics, errors are intermediate between those in applied statistics and in physical chemistry. This is summarized in Table 1.1.

Table 1.1 - Sampling errors

Statistics	Experimental errors	Examples
Traditional statistics	Large	Biometry, Econometrics, Psychometrics
Chemometrics	Intermediate	Clinical chemistry, Environmental chemistry, Industrial process control
Physical chemistry	Small or negligible	Kinetics, Quantum mechanics

(3) *Signal processing*. Most chemometric data are acquired *via* computerized instruments. In most conventional areas of statistics, experimental data is directly measured: for example the length of the root of a plant or the number of pupils in a class constitutes direct raw statistical information. However, because chemists rarely record their data directly, they have a broad range of methods for improving and enhancing the quality of instrumental data. These include Fourier transformation, filtering, instrumental optimization and so on. Better instrumental optimization should result in better analysis. It is, therefore, impossible to separate analysis from acquisition of data, and the chemometrician must take into account signal processing as well as resultant pattern recognition, as part of an integrated strategy for the interpretation and acquisition of laboratory based datasets.

For these reasons, slightly different approaches are required in chemometrics to traditional statistics, although the chemometrician must, nevertheless, base his or her methods on sound statistical principles. In this text we will try to view chemometrics as an essential aid to laboratory experimentation rather than as a branch of applied statistics. Historically, chemometric methods have diverged from applied statistical methods for several of the reasons above. A similar divergence (for different reasons) happened in econometrics in the 1950s, where a variety of techniques were devised to deal with problems inherent in economic data. Econometrics is now regarded, by many, as quite distinct from mainstream statistics.

It is quite likely that the chemometrician of the future will be schooled principally as a chemist, and chemometrics will become a specialized and integrated branch of chemistry like organic chemistry, analytical chemistry, physical chemistry and so on. For the time being, though, the chemometrician relies on methods originally developed by mathematicians and statisticians, and a major job of the chemometrician is to translate these methods into a language and into applications of direct interest to the chemist.

1.2 THE USE OF CHEMOMETRIC METHODS

Chemometrics is a collection of methods for the design and analysis of laboratory experiments, most, but not all, chemically based. Chemometrics is about using available resources as efficiently as practicable, and arriving at as useful a conclusion as possible taking into account limitations of cost, manpower, time, equipment etc. The chemometrician differs from many chemists in that he or she does not always need to understand underlying processes in molecular terms before making conclusions about a system. Indeed, chemometrics often reduces the need for a detailed study of molecular processes. Consider studying the chemical basis of taste. One approach might be to isolate, identify and quantitate each major and characteristic compound in two groups of food that taste different, and try to draw chemical conclusions from this study. There are several difficulties with this approach. The first is that it is extremely time consuming. The second is that it is often hard to isolate and quantitate compounds: it may be necessary to devise a quantitative test for each constituent and if extraction procedures are not perfect (as is usually the case), significant errors will enter into the procedure. The third and most significant difficulty is that it is rarely possible to obtain a complete chemical profile of any natural substance, so only certain compounds will be identified. There is no guarantee that the key compounds have been identified. Obviously, in some well established cases, important chemical constituents of taste are known but this information may have involved many years' work, and is a major undertaking. An alternative approach is to use a statistical method to try to determine the major patterns in chromatograms or spectra of total extracts from different groups of food. The major features can be *calibrated* to taste. This approach will almost certainly produce a good model, probably allowing the chemist to predict the taste of an unknown substance using spectroscopic or chromatographic evidence. What it may not do is allow the chemist to reach a molecular understanding of the basis of taste, but perhaps often such a detailed understanding is unnecessary for the study in hand. Similar examples involve classifying organisms into species and determining the geological origin of sediments. Do we need to build up a full molecular picture of these processes? Inevitably these processes are so complex that our molecular picture is bound to be incomplete and so inaccurate, unless very substantial efforts are made.

Yet most chemists still try to build a molecular description of processes. One of the reasons for this is that traditional chemical measurements tend to be univariate. Until the advent of modern computerized instruments, the ratio between two compounds or the concentration of a single compound in a mixture involved much experimental effort. Calculating the concentration of 10 compounds in a mixture involved major efforts, so many chemists concentrated on one or two marker compounds and made elaborate deductions from the changes in concentration of these compounds. Since there is a large literature in most applications of chemistry, spreading over 50 years or more, chemists are frequently reluctant to change their methods and when first using computerized instruments, these same types of parameters are measured, but instead of calculating the concentration of two compounds manually, spectroscopic peak ratios or similar parameters are used. For example, detailed geochemical kinetic models have been deduced from measurements of the change of ratio between two compounds with geological time. Yet, these models are at best very simple models of natural processes that influence the occurrence of not two but several compounds. Surely the best way to study these processes is by using information about the chemical constituents of the rocks as a whole? It is possible to extract a rock and obtain several hundred peaks using techniques such as chromatography or mass spectrometry. The chemical identities of these peaks may be unknown but does this matter? The question we may be asking is whether there is likely to be oil close to the sediment we are analysing. So long as this question can be answered with a high degree of confidence we do not require a full molecular description of the sediment, or a full understanding of the physical, chemical and biological processes that created the oil.

In many areas of science on the fringes of chemistry, such as pharmacy, environmental chemistry, food chemistry and geochemistry, the use of statistical chemometric models has slowly developed over the last decade probably because of the inherent chemical complexity of these systems. Yet chemometrics still remains to be accepted by the mainstream chemical community, so it is valuable to suggest some examples of how these methods can help in established areas of chemistry.

(1) *Materials science*. Much effort and money is spent developing materials in aerospace engineering that can withstand enormous stress. Generally elaborate physical and chemical understanding of how molecules bind together needs to be reached before new materials are designed and tested. Crystallography, electron microscopy and allied techniques can aid the engineer. Yet statistical approaches, relating hardness to molecular composition are rarely used in this area, and could result in just as economically useful a model of the relationship between structure and property, but in a shorter time.

(2) *Chemical kinetics*. Generally kineticists study single reaction pathways. Yet many natural processes, such as the preservation of food or the burning of a candle, involve several hundred reactions and would be impossible to study by conventional means. Chemometrics can help model these overall reactions statistically without the need to trace each of tens or hundreds of competing pathways.

(3) *Quantum mechanical structure property relationships*. Theoretical chemists traditionally use quantum mechanical calculations to predict molecular properties such as dipole moments, building up a detailed atomic model of each compound in turn. Yet models can equally well be built up by correlating structural parameters to dipole moments in a test series of compounds. Although lacking in molecular insight, a statistical model can probably help predict the dipole moment of a new compound and is easier to establish than a quantum mechanical model.

(4) *Improving organic synthetic yields*. These are influenced by many factors such as pH, temperature, catalyst concentration, solvent composition, relative concentration of reactants etc. Instead of trying to describe the reaction mechanism in molecular terms, often involving a great deal of work, chemometric methods for experimental optimization may be employed.

We will illustrate the need for and use of chemometric methods by two simple examples.

Consider trying to improve the yield of a synthetic reaction. This yield may be dependent on a variety of factors such as, for example, temperature and pH. The organic chemist rarely has the time to produce exact molecular models of how pH and temperature catalyses the reaction so normally uses an empirical approach by trying to perform the reaction at a variety of temperatures and levels of pH. There are several ways of finding this optimum. A simple way might be to set the temperature at a constant value e.g. 40°C and then vary the pH, e.g. over the range 1 to 10, and pick the pH which gives the highest yield. The chemist then keeps the pH constant and varies the temperature until the highest yield is found, and uses this combination of pH and temperature as the "best" values. A graph of yield versus temperature and pH is illustrated in Fig. 1.1.

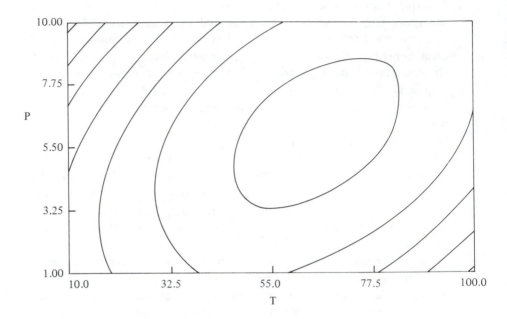

Fig. 1.1 - Contour graph of yield against pH and temperature for a reaction.

A major problem emerges. The graph of temperature versus yield differs according to the value of pH. The graphs for pH 2 and pH 8 are illustrated in Fig. 1.2. The chemometrician says that the factors pH and temperature *interact*, that is, temperature and pH do not influence the yield of the reaction independently. The mechanism of the reaction is likely to differ according to the pH, if the reaction involves acid-base catalysis. Therefore the influence of temperature will differ according to pH values. Equivalently if the yield is very high at one temperature it may be virtually independent of pH at that temperature, whereas at higher or lower temperatures it could show a more pronounced pH profile. Therefore, if we find, first, the optimum pH at a given temperature and then, keeping this pH constant, vary the temperature until the yield is a maximum, we will not necessarily reach a global optimum, as illustrated in Fig. 1.3. In fact, the result will depend on how we planned our experiments and, in part, is a function of experimentation rather than a property of the system. In order to be more sure of reaching a meaningful optimum, a more systematic experimental procedure is required. This could involve testing the yield at 10 combinations of pH (between 1 and 10) and 10 combinations of temperature (between 10 and 100°C). If we want to test all possible combinations, 100 experiments must be performed, involving rather tedious and time consuming efforts. For three factors (e.g. binary solvent composition, temperature and pH)

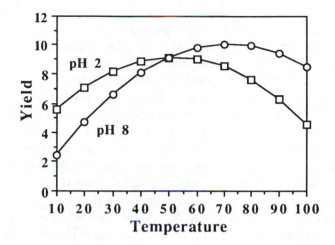

Fig. 1.2 - Graphs of yield against temperature at pH 2 and pH 8 for data in Fig. 1.1, illustrating how the optimum temperature changes with pH.

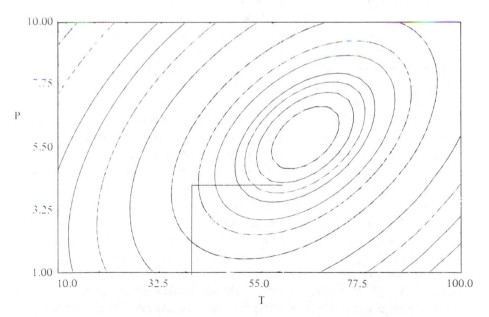

Fig. 1.3 - Attempted optimization strategy for synthetic yield illustrated in Fig. 1.1. The temperature is set at 40^0 and the best pH found at this temperature. Then temperature is optimized at this pH. The final result does not equal the true optimum.

we need to perform 1000 experiments, which is clearly impracticable. Chemometrics involves using systematic approaches to reduce the number of experiments required to estimate this optimum as discussed in later chapters.

Let us assume that we found that, by using a systematic and efficient approach to varying pH and temperature, the true optimum yield was 60%. Yet when we varied pH and temperature, we kept other factors, such as the solvent composition, constant. In many cases yields improve if we add a catalyst. Let us assume that the experiments above were also performed in the absence of a catalyst. If we add some catalyst we expect that yields at all temperatures and pH values will improve, but probably the variation with pH and temperature will be less marked at high catalyst concentration. Probably the overall optimum temperature and pH will change in the presence of a catalyst in the same way that the optimum temperature differed according to pH. So did we locate a true optimum in the absence of the catalyst? Were the experiments in the absence of a catalyst meaningless or a waste of time? The experiment to find the optimum pH at constant temperature did not help us very much so how useful are experiments at a constant (i.e. zero) concentration of catalyst? Should we consider the influence of five or six factors on the yield? Is a two factor experiment useless? There are an infinite number of possible factors that could influence the yield of the reaction, including catalysts and solvents yet to be discovered by mankind. When do we stop experimenting and how has chemometrics helped us?

Chemometrics will help us to be more efficient in our experimentation for equivalent effort by more reliably establishing optimum conditions. For example, instead of performing 100 experiments to establish an optimum combination of pH and temperature, experimental design (discussed in Chapter 2) will normally result in a just as good, or even better, result in about 10 or 20 experiments, reducing the time and effort very considerably. It will also reduce the chance of false optima. Chemometrics, though, is not a total substitute for good laboratory management. The number of factors to be considered, the target yield, how much time we have available and the commercial implications of the work are in the hands of the laboratory manager. There is no single definition of an optimum, nor is there any guarantee that any one method will reach a global optimum. We can, though, take advantage of chemometrics to improve the efficiency with which we work and to improve our chances of obtaining a better answer to a problem within the limitations of the resources at hand.

A second example is that of estimating the proportion of compounds in a mixture. A typical problem involves the estimation of the proportion of chlorophylls *a* and *b* in leaves using uv (ultraviolet) /visible spectroscopy. Normally two characteristic wavelengths are chosen and the ratio of absorbances at these wavelengths are *calibrated* to the chlorophyll ratio. The raw spectra of both pure compounds are also recorded and used to convert the observed ratio at these two wavelengths to a proportion of compounds in the mixture.

Table 1.2 - Simple example of spectroscopic intensities two compounds
monitored at two wavelengths

Compound	Wavelength 1	Wavelength 2
A	1	5
B	3	2

A simple example of this *univariate calibration* is discussed below. Consider recording the spectra of two pure compounds, A and B, at wavelengths 1 and 2. If the intensity of compound C at wavelength i is given by x_{iC} then if the proportion of compound A in a mixture is α and compound B is $(1-\alpha)$, the ratio between spectral intensities at wavelengths 1 and 2 is given by

$$r = \frac{(\alpha x_{1A} + (1-\alpha)x_{1B})}{(\alpha x_{2A} + (1-\alpha)x_{2B})} \tag{1.1}$$

By simple mathematics, Eq. (1.1) can be rearranged to give

$$\alpha = \frac{(x_{1B} - r\,x_{2B})}{(r(x_{2A} - x_{2B})+(x_{1B}-x_{1A}))} \tag{1.2}$$

This is probably best illustrated by a simple numerical example. Consider the spectra in Table 1.2. Using Eq. (1.2), we find that

$$\alpha = (3-2r)/(3r+2) \tag{1.3}$$

We can check that his equation is correct: a pure spectrum of A has a ratio of $r=1/5$. Substituting in Eq. (1.3), we find $\alpha = (3-2/5)/(3/5+2) = 1$. So any ratio of spectral intensities (r) corresponds to a unique value of α.

Many people have used these methods for estimating the relative proportion of two compounds (e.g. pigments) using spectroscopy (e.g. uv/visible spectra) for several years. Yet these methods often yield apparently nonsensical results such as negative concentrations. Why is this? There are several reasons. The first reason is that there is always *noise* or experimental error in a system. So instead of pure compound A giving a ratio $r=1/5$ (=0.2), the ratio will be $(1+\delta_1)/(5+\delta_2)$ where δ is a random noise function. A ratio of 0.15 does not imply a negative amount of compound A, for example, but a slightly noisy spectrum. There are several ways of overcoming this problem. The simplest is by using smoothing functions as described in Chapter 5, which use knowledge of signals and noise to reduce noise levels. Another way is to measure the intensity at several wavelengths: this provides the experimenter with more information, and, hopefully, the noise will average out over these wavelengths. The analysis is now considerably more complex because we are no longer dealing with a single ratio, and methods of analysis such as factor analysis and multivariate calibration, as described in Chapter 6, are employed. A second reason is that a mixture may contain more than two compounds. Although we may be interested principally in chlorophylls *a* and *b*, crude extracts from leaves, for example, will contain several other pigments which will have spectra overlapping, or *interfering* with the chlorophyll spectra. Again chemometric methods will help us disentangle the interferants from the true spectra and so enable us to reach a reliable estimate of the relative proportion of the two pigments.

The amount of time spent developing a method for estimation of the proportion of chlorophylls in a mixture depends, in part, on what resources are at hand. For example if we know that β-carotene is a common impurity we can record the spectrum of pure β-carotene and use this information to subtract for this interferant. How many possible impurities should we consider? Do we have time to synthesize each possible impurity and record its spectrum? How quickly do we need an answer? Will a rough estimate be sufficient or is it worth our while spending six months developing a better method? There is no single answer to the problem, but chemometrics will allow the experimenter to interpret his or her data fairly efficiently. There is no real substitute for chemometric methods for the quantification of multiwavelength spectra of mixtures. But the amount of information and the problems of interferants and noise will vary according to the nature of each application. The laboratory manager cannot automatically pull a chemometrics package off the shelf and expect all his problems to be over. He must carefully assess what resources he is willing to invest in the problem.

In this section we have discussed some possible applications of chemometrics and, hopefully, convinced the reader that it really is worth understanding the basis of chemometrics.

1.3 HOW TO USE THIS TEXT

This text is an introduction to chemometrics. It does not assume a detailed knowledge of statistics and does not discuss computational algorithms in detail. Early chemometrics papers strongly emphasized algorithms because many of the early chemometricians were also good computer programmers. Many of the commonest methods can be expressed in algorithmic form. In particular, matrix operations are easy to program, so many papers emphasize matrix algebra. Although matrices are important for the technically minded, specialized, chemometrician, the majority of users of chemometrics come across the subject through packaged software. They want to be able to interpret and use the output of a computer wisely and safely rather than understand the detailed workings of the relevant software. Some understanding of simple mathematics is also desirable and will, where appropriate, be included in this text, but details of mathematical statistics are left to many excellent texts cited in the bibliographies at the end of the relevant chapters in this book.

Most texts previously available in this field have been written by and for chemometricians. This text is different. It is written for non-chemometricians who need an understanding of how chemometric methods can help in their everyday work. The emphasis in this text is on simple numerical examples supported by extensive graphics. Nearly every example in this text has been programmed into a spreadsheet: the exceptions are very simple calculations which have been performed on a programmable calculator. The interested reader should be able to repeat all the results in this book using most commercially available spreadsheets. Simple multilinear regression, i.e. the ability to fit an observed parameter y, to a variable x, usually using simple least squares straight lines, but sometimes including squared terms (e.g. Section 2.5) should be available. The examples of Section 6.4 and the principal components of the data in Section 7.3 require somewhat more sophisticated calculations but these can still be performed by spreadsheets using the NIPALS and PLS algorithms as described in the appendix: most good spreadsheets have functions to multiply and invert matrices.

All datasets are simulations. It certainly would have been possible to use real data from the literature, but in one or two cases, simple, illustrative datasets are not available so for the sake of consistency we use simulations in all examples. The reader should refer to the bibliographies at the end of each chapter for further examples.

It is possible to perform most chemometric calculations on a variety of commercial software. These range from sophisticated statistical packages such as SAS, S, SPSS and GENSTAT to specialized chemometrics software such as the SIMCA package. There is also a growing interest in chemometrics software from instrument manufacturers, but much of this dedicated on-line instrumental software is fairly restricted and aims to be user-friendly rather than comprehensive. It is possible to program most common statistical and chemometric methods into languages such as FORTRAN but this approach lacks extensive graphics output and extensive input-output routines: about 90% of specialized chemometric software involves a user interface because the casual user is unlikely to have the time or interest to edit FORTRAN programs, and chemometric results are best presented graphically.

There is no universal software package that solves all the problems of chemometrics, hence the diverse market. The best approach depends on your background and interest. For example, highly user-friendly and packaged instrumental software is very suitable for technicians who are not too interested in the statistical basis of results but can frustrate the more mathematically minded user. At the other extreme packages such as GENSTAT provide excellent and comprehensive statistical facilities but are not user-friendly and do not have comprehensive graphical facilities. Packages intermediate between these extremes such as SAS do include comprehensive graphics facilities and a wide range of non-statistical features but are quite hard for the non-computer expert to use: it is possible to produce SAS macros and customize particular applications to a specific purpose, but there must be one expert in the group. In addition to the software package, it is important to choose the machine on which to use the software. Micros such as Apple Macintoshes are useful for simple tutorial examples. Extensive calculations may need to be performed off-line using a VAX or SUN computer, or even a larger mainframe. Problems of data transfer from one computer to another, often involving special problems of decoding data from one format to another, must be dealt with.

It is important, though, that the reader of this book has access to some software. Because each laboratory has different packages this book has not been tailored to any specific approach. It is, however, hard to appreciate the possibilities of chemometrics without hands-on use of software.

Finally this book is not meant to be a substitute for or dependent on having attended a general statistics course. There is a well established area of statistics for analytical chemists covering topics such as the precision and accuracy of measurements. There is inevitably some overlap between general statistics and chemometrics but the reader is referred to books such as *Statistics for Analytical Chemistry* by Miller and Miller for a general introduction. General statistical tables may be useful to the reader but in this book we only refer to the normal distribution, F-test and χ^2-test. It is not necessary for the reader to understand the detailed theory behind these distributions and tests, although this information is available in most specialized statistics books. It is also possible to obtain these distributions from most numerical software packages.

1.4 THE LITERATURE OF CHEMOMETRICS

Some readers of this text will want to delve into the more specialized literature of chemometrics.

The biennial reviews in the journal *Analytical Chemistry* are the most regular literature surveys of chemometrics available and point the reader the main primary publications in the area. These reviews do not aim to be comprehensive and concentrate primarily on work of interest to the analytical chemist. The review of chemometrics in *The Analyst* in late 1987 is a more critical review of the subject but does not aim to cover the literature as broadly.

There are two journals, namely, *Chemometrics and Intelligent Laboratory Systems* (Elsevier) and *Journal of Chemometrics* (Wiley) that are of interest to the chemometrician. The former journal is slightly more applied than the latter journal, although both regularly report new statistical and computational methods. In addition, *Chemometrics and Intelligent Laboratory Systems* publishes regular tutorial papers which involve explanation and discussion of both new and well established methods, mainly illustrated by various examples. The collected reprints of the tutorial papers of the first five volumes of *Chemometrics and Intelligent Laboratory Systems* are published by Elsevier. Various other journals publish occasional short tutorial articles on chemometrics: these include *Analytical Proceedings* and *Trends in Analytical Chemistry*. In the UK, the UK Chemometrics Discussion Group publishes a regular newsletter three or four times year year which is available to all members (who do not need to live in the UK). Some of the early chemometrics papers appear in *Analytica Chimica Acta, Computer Technology and Optimization* section.

The applications of chemometrics are spread widely around the literature. Examples of journals in which occasional chemometrics papers appear include *Analytica Chimica Acta, Analytical Chemistry, The Analyst, Journal of the Chemical Society Perkin Transactions II, Tetrahedron Computer Methodology, Environmental Pollution, Journal of Chemical Information and Computer Science, Journal of Computer Aided Molecular Design* and *Journal of the Royal Statistical Society*, to name only a few that this author regularly refers to.

There are several books of interest to chemometricians, most of which are referenced in the relevant chapters. There are two existing overviews of chemometrics. The text by Sharaf *et al*. was the first comprehensive text with chemometrics in its title and is a good reference for the mathematically minded analytical chemist. The textbook entitled *Chemometrics : a textbook* by Massart *et al*. is aimed more at the laboratory manager, and is less mathematical. Both texts, should be in all good chemometrics libraries. The collection of papers from the Cosenza NATO Advanced Study Institute in chemometrics held in 1983, edited by Kowalski (*Chemometrics : Mathematics and Statistics in Chemistry*), was an historical milestone in chemometrics. Although most of the material is superseded, it provides a good selection of source material and most of the prominent early workers in this field contributed articles to the book. There have been two notable ACS symposium volumes on chemometrics. A series of monographs published by Research Studies Press jointly with Wiley cover some areas of chemometrics in detail.

There are several general texts on statistics applied to other scientific disciplines. These include the book by J. C. Davis on *Statistics and Data Analysis in Geology* and various texts on statistics aimed at biologists, including the text on *Mathematics and Statistics for the Biosciences* by G.Eason *et al*. Texts on general statistics are far more common in geology and biology libraries and bookshelves than chemistry libraries and bookshelves, although with the widespread interest in chemometrics, chemists are likely to become more interested in statistics in the near future, so this will probably change. Not all the people that contributed to the early development of chemometrics thought of themselves primarily as chemometricians, and it is worth noting the texts by Massart *et al.* on *Evaluation and Optimization of Laboratory Methods and Analytical Procedures* and by Kateman and Pijpers on *Quality Control in Analytical Chemistry* both of which cover many topics relevant to chemometrics. Several books on pattern recognition and on signal processing are of general interest and are referenced in the relevant chapters.

REFERENCES

Reviews
B.R.Kowalski, *Anal. Chem.*, **52**, 112R (1980)
I.E.Frank and B.R.Kowalski, *Anal. Chem.*, **54**, 232R (1982)
M.F.Delaney, *Anal. Chem.*, **56**, 261R (1984)
L.S.Ramos, K.R.Beebe, W.P.Carey, E.Sanchez, B.C.Erickson, B.R.Wilson, L.E.Wangen and B.R.Kowalski, *Anal. Chem.*, **58**, 294R (1986)
R.G.Brereton, *The Analyst*, **112**, 1935 (1987)
S.D.Brown, T.Q.Barker, R.J.Larivee, S.L.Monfre, H.R.Wilk, *Anal. Chem.*, **60**, 252R (1988)
S.D.Brown, *Anal. Chem.*, **62**, 84R (1990)

Books
M.A.Sharaf, D.L.Illman and B.R.Kowalski, *Chemometrics*, Wiley, New York, 1986
D.L.Massart, B.G.M.Vandeginste, S.N.Deming, Y.Michotte and L.Kaufman, *Chemometrics: a textbook*, Elsevier, Amsterdam, 1988
J.C.Miller and J.N.Miller, *Statistics for Analytical Chemistry, Second Edition*, Ellis Horwood, Chichester, 1988
B.R.Kowalski (Editor), *Chemometrics: Mathematics and Statistics in Chemistry*, Reidel, Dordrecht, 1984
B.R.Kowalski (Editor), *Chemometrics: Theory and Application*, ACS Symposium Series No. 52, American Chemical Society, Washington, DC, 1977
D.A.Kutz (Editor), *Chemometrics Estimators of Sampling, Amount and Error*, ACS Symposium Series No. 284, American Chemical Society, Washington, DC, 1985
J.C.Davis, *Statistics and Data Analysis in Geology, Second Edition*, Wiley, New York, 1986
G.Eason, C.W.Coles and G.Gettinby, *Mathematics and Statistics for the Bio-Sciences*, Ellis Horwood, Chichester, 1980
D.L.Massart, A.Dijkstra and L.Kaufman, *Evaluation and Optimization of Laboratory methods and Analytical Procedures*, Elsevier, Amsterdam, 1978
G.Kateman and F.J.Pijpers, *Quality Control in Analytical Chemistry*, Wiley, New York, 1981

2

Experimental design

2.1 WHY DESIGN EXPERIMENTS?

Probably the first stage in laboratory studies is to plan some experiments. These may involve sampling river water to see whether any toxic chemicals are present, improving the yield of a synthesis, observing the spectra of a series of solids under different conditions and so on.

In a systematic world we sit down together, formulate a problem and work out how best to tackle it experimentally. Chemists have been doing this for decades, and so there is often strong resistance to changing age old methods of working and employing statistical approaches for experimental design, since these differ, in many cases, from conventional chemical wisdom. Perhaps the first difficulty is a psychological one. Why should chemists use statistical methods for experimental design?

Possibly the first limitation with traditional chemical approaches is that they are *univariate*. Simple physical laws can be derived where one, two or at the maximum three well behaved *factors* (independent parameters) influence the measurable behaviour of a chemical system called a *response*. At an early stage of a student's training, teachers and lecturers go to great lengths to design experiments that, apparently, *work*. That is they design apparently foolproof experiments. If the result of the experiment is supposed to be a linear relationship, then great effort is taken to build and maintain apparatus that actually behaves in a linear fashion. This is deeply culturally engrained in the minds of many chemists. This means that chemists work in an apparently systematic fashion, first investigating one effect, then another and so on. Investigating too many interacting factors at the same time seems too complex, and indeed exact physical models often break down under such circumstances.

In the real world, most natural processes are the result of many complex interacting factors, and if simple laboratory experiments are to be extended to the outside world, it is vital to understand and model these complex processes. An example cited in Chapter 1 was the dependence of the yield of a chemical reaction on pH, temperature, catalyst concentration and so on. The analysis was really quite difficult, and we got into a number of intellectual tangles whilst trying to discover what really was the "best" answer. Statistical approaches will help here.

But experimental design can also help in simple quantitative experimentation. In order discuss this further we need some definitions.

(1) A variable or *factor* is changed throughout an experiment. This might, for example, be pH or temperature. The factor is generally under the control of the experimenter. It is normally denoted by x.

(2) A *response* is observed from each experiment. This response might be the yield of a reaction or the chromatographic separation of two peaks. This is not under the direct control of the experimenter, who is interested in how the response depends on the factor.

(3) Several experiments are performed in order to determine the relationship between the response and the factor. We will define the total number of experiments as N, so that the value of the response for experiment n is given by y_n and the value of the factor by x_n.

(4) There is no reason to restrict experiments to a single factor, single response system. Multifactorial experiments are common in chemistry and will be discussed in detail later in this chapter: the yield of a reaction might, for example, depend on both pH and temperature in which case the experiment is a two factor experiment. Multiresponse experiments are also common: for example, temperature and pH might influence the resolution of several chromatographic peaks. Although there is no universal notation we will denote the value of the ith factor in the nth experiment by x_{in} and use a similar convention for the response. The total number of factors will be denoted by I.

The objective of a single factor, single response experiment might be to determine whether y is linearly related to x. A common way of testing linearity is by calculating the *correlation coefficient* which is defined by

$$r_{x,y} \quad = \quad \frac{\sum\limits_{n=1}^{n=N}\{(x_n-\bar{x})(y_n-\bar{y})\}}{\sqrt{\{\sum\limits_{n=1}^{n=N}(x_n-\bar{x})^2\}\{\sum\limits_{n=1}^{n=N}(y_n-\bar{y})^2\}}} \qquad (2.1)$$

where $\bar{x}$ and $\bar{y}$ are the means of x and y respectively, and the other parameters are as defined above. If this coefficient equals ±1 the two parameters are said to be perfectly correlated, and this suggests a linear relationship. As can be verified by simple statistics and graphics, correlations close in magnitude to 0 suggest poor correlation.

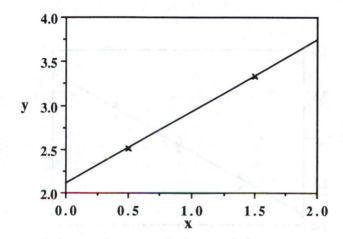

Fig. 2.1 - Correlation between two variables : two unreplicated experiments.

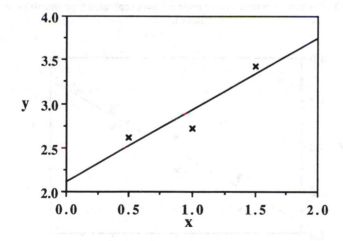

Fig. 2.2 - Correlation between two variables : three unreplicated experiments.

Consider the nature of the experiment used to measure the correlation coefficient. A simple approach might be to record two values of y at two different values of x (Fig. 2.1). This experiment *must* lead to a correlation coefficient of ±1, whether or not the two variables are related. The experiment is poorly designed. Whatever the true underlying trend, the same quantitative answer is obtained. Consider a slightly better experiment, involving three independent measurements (Fig. 2.2). There is now some information as to whether the parameters are linearly related, but this information is still rather

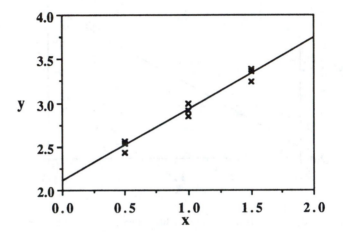

Fig. 2.3 - Correlation between two variables : three replicated experiments (low noise).

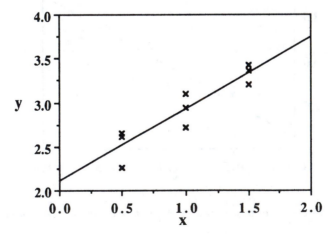

Fig. 2.4 - Correlation between two variables : three replicated experiments (high noise).

limited. When do we stop experimenting? In Chapter 1 we discussed the importance of considering the errors or noise in the system. Fig. 2.3 represents a highly reproducible experiment, whereas Fig. 2.4 illustrates a less reproducible experiment. The information as to reproducibility of an experiment is provided by taking *replicates*.[1] If, for example, we recorded four non-replicated points and the correlation coefficient was low in magnitude we would be justified in suggesting that the relationship was non-linear if the analytical (replicate) error is small as in Fig. 2.3. If this experimental error is high, then a small correlation might be a consequence of experimentation and not necessarily indicative of an underlying non-linear relationship. Consider, for example, an experiment to measure the build up of a toxic chemical in fish in a river, after a spillage. Does the concentration of toxin increase linearly with time? We may want to construct a graph of toxin concentration per unit weight of fish tissue *versus* time. There may, however, be several ways of measuring this concentration. One might be by gas chromatography-mass spectrometry (GC-MS) which is difficult to reproduce quantitatively. Another approach might be by gas chromatography (GC) alone, which is far more reproducible quantitatively, but contains less diagnostic information. Thus we expect greater replicate errors using GC-MS, so a low correlation coefficient does not necessarily suggest that there is a lack of linearity between time and toxin concentration. On the other hand a low correlation coefficient obtained in a GC experiment is more *significant* than in a GC-MS experiment. This means that a low GC correlation coefficient really does imply that there is very little evidence for linearity.

There, in a test as to whether a response (*y*) is linearly related to the value of a factor (*x*), both the number of experiments and the number of replicates provide important information. A numerical result such as a correlation coefficient can only be interpreted in the light of the experimental design. Clearly if 50 independent experiments are performed and there is still a correlation coefficient close to ±1, it is very likely that the two variables are linearly related, but we may still prefer some numerical guidelines as to how well the experiment is going. Statistics can aid us here.

There is yet another use for experimental design and that is the study of *interaction effects*. These were briefly mentioned in Chapter 1 when discussing synthetic yields. We commented that the optimum values of pH and temperature differed according to catalyst concentration and so the influence of the three factors pH, temperature and catalyst concentration on the observed response (reaction yield) are not completely independent. That implies that these factors *interact*. So in a laboratory study, the influence of an isolated factor on a

---

[1]This has a formal statistical meaning. The response(s) is (are) measured for identical values of the factor(s) i.e. *y* is measured several times whilst keeping *x* constant.

response may not necessarily correspond to the influence of the same factor in a *multivariate* situation. Often interaction effects are presented as of rather obscure statistical interest. However some estimate of the size of such effects can be vital for the proper interpretation of laboratory experiments. Consider, for example, an experiment in which unicellular marine algae are grown at varying temperatures and the concentrations of various secondary metabolites are measured. It might be possible to find a function of these concentrations that is proportional to temperature. Perhaps, we then want to take a sample of marine algae (this is a very common experiment in geochemistry, where, of course, the algae are dead but the fossils still contain chemicals), and predict the temperature at which the algae were living. Normally highly controlled and reproducible apparatus is built in the laboratory in order to study carefully the influence of change of temperature without unpredictable changes in other conditions such as light, growth environment and so on. Yet nature is not so kind, and other factors may also vary simultaneously. Is the relative concentration of secondary metabolites also affected by these other factors? We need to know this before interpreting the field data and this can be done by designing careful experiments to estimate the interactions between these various factors. There are, of course, many other cases where experiments are virtually useless without some knowledge of these effects, and many classical examples where predicted behaviour based on simple model systems differs significantly from observed behaviour of complex systems because of the failure to take interactions into account.

To summarize there are three main reasons for statistical design of experiments.

(1) *Quantitative modelling*. This is probably the commonest reason. The quantitative influence of one or more factor(s) on one or more response(s) is one of the commonest reasons for scientific experiments. An example is the determination as to whether there is a linear relationship between x and y.

(2) *Optimization*. This has particular significance in chemistry. Finding the conditions that provide the best yield in a reaction or that best resolve peaks in a spectrum are examples.

(3) *Quality control*. We will not discuss this in detail in this chapter. There are modern statistical approaches such as Taguchi methods which have been used, for example, to increase industrial productivity.[1]

Experimental designs should take the following into consideration.
(1) The overall number and arrangement of experiments.

---

[1]There is some difference in terminology between analytical chemists and statisticians in this area. Some of the approaches used by analytical chemists to monitor the quality of processes are described in Chapter 3.

(2) The number of replicates required to obtain knowledge about the reproducibility of the system.

(3) The interaction effects.

2.2 CHOOSING A DESIGN

The first step in experimentation should be to establish the nature of the problem and define the required product of the experiment. With this knowledge an appropriate design can be chosen. It is important to think clearly about the reason for experimentation, what is known about a system, and what problems may be encountered.

There is little point performing complex statistical analysis if we ignore basic and obvious intuitive knowledge. The first question to ask is normally whether the experiment is one of optimization or modelling. If the main desire is to optimize a system (e.g. to resolve two isomers in an HPLC (high performance liquid chromatography) experiment) is there any point building a quantitative model of the system? Also, how far is it necessary to optimize the system? If the main objective of the experiment is to separate two isomers fully, for example, once this has been achieved, further optimization is of no interest. In other cases, such as optimizing the yield of a reaction, optimization is almost always likely to achieve a better answer so it is worth continuing until a true optimum has been found. In other cases some criterion of quality may need to be built into a system (e.g. in chromatography we might want to continue experimenting until a resolution between two peaks is of a given quality: although further experiments will continue to improve resolution the desired quality has been reached so there is no point in further optimization).

Modelling experiments are different in nature to optimization experiments since quantitative parameters are required from the experiment and so a good knowledge of the system is needed. Under such circumstances it might be useful to say that there is, for example, 95% likelihood that a given model is obeyed. In order to do this, it is first necessary to propose a model. Normally this is in the form of a polynomial

$$\hat{y} = b_0 + \sum_{i=1}^{i=I} b_i x_i + \sum_{i=1}^{i=I} b_{ii} x_i^2 + \sum_{i=1}^{i=I-1} \sum_{j=I-i+1}^{j=I} b_{ij} x_i x_j + \sum_{i=1}^{i=I} b_{iii} x_{iii}^3$$

$$+ \quad .... \tag{2.2}$$

where $\hat{y}$ is a predicted response, there are I factors, b_0 is an intercept term (the mean of all the experimental responses, in fact), and the other coefficients are normally determined experimentally. The values x_i are the *values* of each of the I factors. These are related to the actual values of the factors (e.g. concentration of a chemical) by a process of *coding* which is discussed in detail below. For a single factor experiment this reduces to

$$\hat{y} \quad = \quad b_0 + b_1 x + b_{11} x^2 + b_{111} x^3 + \quad \ldots\ldots \quad (2.3)$$

which is a simple polynomial in x. It is normal to restrict the model to a certain power of x, for example, a quadratic. Eq. (2.3) arises from a *single factor* experiment. Most responses are dependent on several factors. A *three factor quadratic model* is given by

$$\hat{y} \quad = \quad b_0 + b_1 x_1 + b_2 x_2 + b_3 x_3 + b_{11} x_1^2 + b_{22} x_2^2 + b_{33} x_3^2 +$$
$$b_{12} x_1 x_2 + b_{13} x_1 x_3 + b_{23} x_2 x_3 \qquad (2.4)$$

In Eq. (2.4) we see that there are three cross-product terms of the form b_{ij} which correspond to the interactions as discussed above.[1] The objective of an experiment may be formulated statistically as deducing a quantitative model of the form of Eq. (2.4). Normally regression methods are used to determine the coefficients b from the observed response y at different levels of x. A good experiment should provide enough information to estimate these coefficients reliably, and quantitative information as to how well these coefficients have been estimated. The form of the model is important and dictates the minimum number of experiments required. If the model consists of too few terms, then the experiments might not properly describe the system. If the model consists of too many terms, many more experiments than are necessary would be performed. There clearly needs to be some sensible assumptions about how elaborate a model is required.

Another common difficulty encountered when designing experiments is in the estimation of interaction terms. Consider a five factor experiment. A quadratic model, such as the one derived above, will have 21 terms namely
(1) an intercept term,
(2) five linear terms,
(3) five squared terms for pure factors,
(4) 10 two factor interaction terms.

Are all interactions significant? Is there likely to be an interaction between factors 1 and 5? If all interactions are to be tested, then the number of experiments normally becomes prohibitively large, so a consideration of which factors are most likely to interact can keep the size of large experiments in proportion. Higher order interactions such as three factor terms and straight one factor cubic terms will increase the number of experiments still further, so some means of using knowledge of the system to reduce the number of experiments is necessary, particularly since many higher order interactions are unlikely. This will be discussed in more detail in Section 2.4.

---

[1] Note that there is another possible interaction term between these three factors, namely one of the form $b_{123} x_1 x_2 x_3$ but that this term corresponds to a cubic rather than quadratic model.

Once the objective of the experiment has been decided upon, various aspects of experimentation must be considered. The first and most important is the choice of factors. Obviously, the more factors chosen, the more useful and complete the experiment. However, more factors involve more work: some intuitive estimate of which factors are most important should be made. Another vital consideration is the relative influence of the factors on the experiment. For example, if the objective of an experiment is to optimize the separation of isomers chromatographically using HPLC, are pH, temperature, flow rate, solvent composition and so on equally crucial, and over which region does changing these factors most influence the response? If one factor is far more influential than another, then the raw values of the factors should be *coded* to take this into account. If a sensible pH range is between 4 and 8, whereas a sensible temperature range is between $40^\circ C$ and $80^\circ C$, it might be sensible to perform a mathematical transformation so that the upper and lower sensible temperature and pH bounds correspond to similar values of x. This coding is often performed by one of the following transformations.

$$x_i' \quad = \quad 2c \, \frac{(x_i - (x_{ihi} + x_{ilo})/2)}{(x_{ihi} - x_{ilo})} \tag{2.5}$$

or (for a logarithmic scale),

$$x_i' \quad = \quad 2c \, \frac{\log(x_i/(\sqrt{x_{ihi} \times x_{ilo}}))}{\log(x_{ihi}/x_{ilo})} \tag{2.6}$$

where x_{ihi} and x_{ilo} are chosen to be the sensible upper and lower limits for a given x (e.g. pH 4 = x_{ilo} and pH 8 = x_{ihi}) and c is a constant. If we want to scale x_{ilo} and x_{ihi} to ± 1, then we set $c=1$. There are, of course, other possible methods for scaling, but linear or logarithmic transformations are normally the most straightforward. An example of this scaling is provided in Table 2.1. This scaling is quite important. For example within the range for pH and temperature cited in Table 2.1, we might want a change in temperature of $10^\circ C$ to be of equal significance to a change in pH of 1. If the values were not properly scaled a change in temperature of $1^\circ C$ might appear to be as significant as a pH change of 1, so we would spend most of the time fine-tuning the temperature and only using a coarse search for the pH. Many parameters such as temperature, pH, solvent concentrations etc. are measured on very different absolute scales so that we need to convert to a statistically convenient scale. Obviously this coding is somewhat arbitrary but must be considered prior to experimentation. A logarithmic scale allows for physically meaningful negative values and is frequently useful. The response can also sometimes be coded. For example, should an increase in yield between 50% and 60% be as significant as an increase between 95% and 96%? A form of coding that set $y' = \log(100-y)$ would take care of this. The most sensible method of coding depends on the experiment under consideration.

Table 2.1 - A possible scaling of experimental variables

Variable	High	Intermediate	Low
pH			
Actual	9	7	5
Coded	+1	0	−1
Catalyst concentration (mM)			
Actual	100	10	1
Coded	+1	0	−1
Temperature (°C)			
Actual	80	60	40
Coded	+1	0	−1

Note : catalyst concentration is scaled logarithmically, whereas pH and temperature are scaled linearly.

The next consideration is whether any unforeseen problems are likely to occur during the experiments. A set of experiments might lie half completed whilst an apparatus malfunctions. Can something be obtained from the preliminary experiments, or is it essential to complete the entire set of experiments? This can normally be taken into account by the order in which the experiments are performed. Although the result of a design might be a table of conditions, running the experiments in a systematic way in which conditions are varied from beginning of the table to the end often requires the entire analysis to be complete before any meaningful results are obtained. Consider a three factor experiment. It is possible to design the order in which the experiments are carried out so that the influence of two factors (often chosen to be the most significant ones) can be estimated from the first few experiments and then the later experiments build up a picture of the influence of the third factor.

Further problems concern long term drift in performance of apparatus and biological material. If experiments are to take several months to run, it is important to build in checks and replicates of the system throughout the experimental period. Random designs are sometimes used in which the order of the experiments is randomized. For example if we want to understand how the rate of a chemical reaction varies with temperature and decide to run the reaction at five temperatures between 0°C and 100°C it is best that we do not run experiments at successively increasing or decreasing temperatures in case other factors influence the performance of the apparatus with time. Obviously we hope that these extraneous factors do not adversely influence the results but it is not always easy to predict in advance.

Finally the chemometrician must choose between *simultaneous* and *sequential* designs. In the former several experiments are performed, often, but not necessarily, simultaneously. After considering the objective of a series of experiments, a set of conditions is established, usually displayed in tabular form. These conditions might be of the form: Experiment 1, use concentration A, temperature B and pH C, etc. All experiments need to be carried out. The resultant data are in the form of a list of responses, which can then be modelled, from which a good description of the variation of response with factors is usually obtained, which can then be used for optimization, further modelling, quantitative prediction etc. This approach is normally the most informative but does frequently involve quite a number of experiments. Sequential experiments are generally only useful for optimization studies. Experiments are successively performed until a desired optimum is reached; the experimental design guides the user as to which experimental conditions should be used next and whether a final solution has yet been found. It is not possible to obtain a quantitative model of the response, but this approach can be useful when sequential experiments are mandatory and there are no facilities for regression modelling of the response afterwards. For example, the simplex approach (see Section 2.3) has proved popular for tuning of scientific instruments. A peak (e.g. the width of an nuclear magnetic resonance (NMR) signal) is monitored as various magnetic field controls are modified. A search is made of the direction in which the width of the peak is minimized, and the conditions provided by the simplex are regarded as the best conditions for tuning the instrument. It is only possible to take one spectrum at a time, the answer is required fairly rapidly, and it is of no spectroscopic interest how the magnetic field homogeneity varies with field controls.

There are several theoretical texts on experimental design to which the interested reader is referred in the bibliography. Most of these have been written by statisticians and reflect the cultural difference between statistics and chemistry. The majority of classical examples in areas such as biometrics involve problems where each experiment is expensive, so the field scientist might consult a statistician in advance, spend several days taking each measurement, and then return with a pile of data which the statistician will analyse at length and attempt to draw quantitative conclusions from. These problems are very different from many chemical ones in which all the chemist wants to do is to separate some compounds, tune an instrument or optimize the yield of a synthesis and so has little time or interest in the theoretical modelling of the response, and many of these experiments can be performed rather rapidly on-line. Even if the chemist throws away some statistically interesting information this is often of less interest than achieving reasonable, although statistically suboptimal, and sometimes unsound, results in a short time. One of the most developed areas of chemical experimental design is in chromatography, especially with reference to the pharmaceutical industry: separations of synthetic impurities, isomers and breakdown products from a preparative reaction involve choosing chromatographic conditions. The approach adopted by chromatographers is in many ways completely different to that of the statistician, but behind all these applications of experimental design are the problems of interactions, replicates, achieving as much as possible in as short a time as possible and so on. Experimental design has been used less for quantitative modelling in chemistry but is equally applicable, although some appreciation of matrix algebra and classical mathematics is required for the use with quantitative model building. It is, therefore, important to evaluate the magnitude of the problem under study. Is the problem such that a quantitative knowledge of the response is required or can a simple and rather unrigorous optimization suffice? Many statisticians might feel that several applications of experimental design in chemistry are unrigorous and poorly thought out, but it is vital to appreciate when formal methods are required and when a less statistically sound approach is adequate. Chemometrics is very different to traditional statistics in which a quantitative model of a system is almost always required. Laboratory based experimenters often have very different needs and they must take these into account when choosing a method.

2.3 SEQUENTIAL METHODS

The most well publicized sequential design in chemistry is the *simplex* method. This approach can probably most easily be understood by reference to a single factor experiment. Consider optimizing the process illustrated in Fig. 2.5, where the observed response is rate of a reaction as pH (the experimental factor) is varied. The reaction might be an acid catalysed reaction, since the maximum rate is observed at pH 4. The experimenter is likely to try running the reaction at various pH levels, since he or she will not know the pH rate profile in advance. For example, let us assume that the experimenter tries, initially, to run the experiment at pH 8 and pH 7. The rate at pH 7 is faster than that at pH 8. So probably the reaction will be still faster at even lower pH. The experiment at pH 6 yields a faster rate and if the investigator then reduces pH systematically by steps of 1, the rate will increase until pH 4 is reached and then decrease afterwards. The results of the experiment are tabulated in Table 2.2.

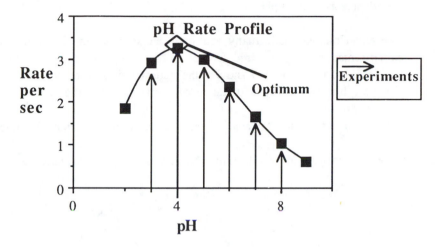

Fig. 2.5 - pH rate profile experiment.

Table 2.2 - Results of the pH rate profile experiment

Experiment number	pH	Rate (s^{-1})
1	8	1.05
2	7	1.65
3	6	2.35
4	5	2.98
5	4	3.26
6	3	2.90

   The experimenter is justified in picking pH 4 as the *optimum* value of pH. However in order to find this value, the experimenter used various rules for experimentation, which might be summarized as follows.

(1) Choose two starting pHs, (7 and 8), and record the the rate at these pHs. The "distance" between these values of pH is 1 pH unit and is called the *step-size*.

(2) Work out which pH is best (i.e. higher rate) and test 1 pH unit away in the direction of the previous best pH.

(3) If the new pH gives an even better rate constant, continue with the search.

(4) If the new pH gives a worse rate constant, stop at the previous pH, which is
    adopted as the best pH.

The experimenter made many arbitrary decisions at the beginning of the
experimentation.
(1) He arbitrarily chose pH 7 and 8 as the starting points of the optimization.
(2) He arbitrarily chose a distance of 1 pH unit for the "moves" in the
    experiment.
(3) He arbitrarily scaled the acidity factor and used, for example, pH rather than
    $[H^+]$.
(4) He arbitrarily restricted the moves to a fixed size (i.e. 1 pH unit).

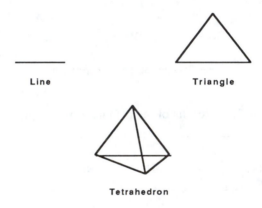

Line   **Triangle**

Tetrahedron

Fig. 2.6 - One, two and three dimensional simplex figures.

Often these decisions were probably made sensibly and with some advance
knowledge of the system. This approach is can be called *one dimensional
simplex optimization*. A simplex is defined as the simplest possible $p-$
dimensional geometric figure.[1] In one dimension this is a line, but in two
dimensions it is a triangle and in three dimensions a tetrahedron. These are
illustrated in Fig. 2.6. Each dimension, or axis, corresponds to a coded factor.
The experimenter performs experiments at the corners of the simplex, so in the
one factor case discussed above, the line is of length 1 pH unit, representing
two experiments (initially at pH 7 and 8). The graph of response against the
value(s) of the factor(s) is often called a *response surface*. In two dimensions
this can be quite complex and because of interactions between variables not

---

[1]A simplex can also be called an $p-$dimensional hypertriangle. It is a figure bounded by $p+1$
sides all of which are equal in length.

always simple to search. As explained in Chapter 1, the approach of first
optimizing one factor and then the next does not always yield a true optimum,
so a more systematic approach is required. In Fig. 2.7 we illustrate a two factor
response. How can we adapt the simplex approach to finding the optimum of
this response surface?

A two dimensional simplex is a triangle, so after suitably scaling the factors,
as discussed above, three (= $p+1$) starting points should be chosen: note that
when analysing the one factor experiment above we chose only two starting
points. We should start with an equilateral triangle, so that each side is equal in
length: each corner of the triangle provides us with experimental conditions. Of
course, we can always scale the dimensions so that each side corresponds to a
different real change in value of a parameter as discussed in Section 2.2. If two
co-ordinates are $(x_1, x_2) = (0, 0)$ and $(0, 1)$, then the third co-ordinate is $(0.5,
0.866)$, as can be verified by simple geometry (these are the co-ordinates of an
equilateral triangle). The length of each side of the triangle is the *step-size*. The
responses at all three vertices are then recorded (note we record $p+1$ responses
for a p factorial problem). The worst response is then discarded, in the case in
question a response at $(0, 0)$. The next experimental conditions are calculated as
follows. A new equilateral triangle is constructed consisting of the best and next
best response [i.e. those at $(0, 1)$ and $(0.50, 0.866)$] and a new point, which is
the reflection of the worst response in the line formed by the other two
responses. The calculation of this new point is performed as follows.

(1) The three initial responses should be ranked, so that the best response is
    labelled 2, the next best is labelled 1 and the worst 0.
(2) The co-ordinates of the best response are given by $x_2 = (x_{12}, x_{22})$ and the
    next best by $x_1 = (x_{11}, x_{21})$, and the worst by $x_0 = (x_{10}, x_{20})$, where x
    represents a vector.
(3) The co-ordinates of the new point are given by
$$x_{new} = (x_{1new}, x_{2new}) = [(x_{12} + x_{11}) - x_{10}, (x_{22} + x_{21}) - x_{20}] \quad (2.7)$$

It is easiest to illustrate this with a simple numerical example. If responses of
the three experiments above are ranked so that $x_0 = (0, 0)$, then the value of
$x_{new} = ((1 + 0.5) - 0, (0 + 0.866) - 0) = (1.5, 0.866)$.

Once a new point is calculated the response at this point is computed. The
new simplex now consists of points x_1, x_2 and x_{new}. In the jargon of simplex
optimization we have *discarded* x_0 and kept x_1 and x_2. If the new point is better
than one of the two previous points x_1 and x_2, the calculation is continued, and
yet another response is computed. The best and second best points are used
again to calculate a new point. If the new point is worse than either of the best
and second best points then return to the original points and calculate

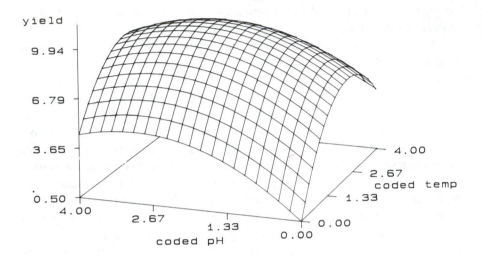

Fig. 2.7 - A two dimensional response surface.

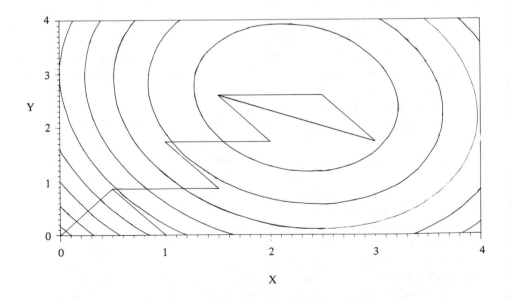

Fig. 2.28 - Graph of *h* for face centred cube (upper right and upper left curve)
and central composite (lower right and lower left curve) designs for one factor
where the other two factors are held at coded values of 0.

$$\mathbf{x_{new}} = (x_{1new}, x_{2new}) = [\,(x_{12} + x_{10}) - x_{11}, (x_{22} + x_{20}) - x_{21}]\quad (2.8)$$

instead. The point $\mathbf{x_1}$ is discarded, and $\mathbf{x_{new}}$, $\mathbf{x_2}$ and $\mathbf{x_0}$ are used in the new calculation. If the new point is still the worst, then it is likely an optimum has been found.

Table 2.3 - Fixed size simplex optimization

Move	Co-ordinates		
0	**0 0** **0.5**	**1 0** **4.5**	**0.5 0.866** **4.886**
1	**1.5 0.866** **7.713**	1 0 4.5	0.5 0.866 4.886
2	1.5 0.866 7.713	**1 1.732** **7.850**	0.5 0.866 4.886
3	1.5 0.866 7.713	1 1.732 7.850	**2 1.732** **9.503**
4	**1.5 2.598** **9.390**	1 1.732 7.850	2 1.732 9.503
5	1.5 2.598 9.390	**2.5 2.598** **9.870**	2 1.732 9.503
6	**3 1.732** **9.157**	2.5 2.598 9.870	2 1.732 9.503
7	**1.5 2.598** **9.390**	2.5 2.598 9.870	2 1.732 9.503
8	3 1.732 9.157	2.5 2.598 9.870	2 1.732 9.503

Note : co-ordinates and response for the three points in each move of the simplex optimization are tabulated. The new point and response is displayed in bold for each move. The optimum found by the simplex procedure is outlined in shadow format.

Occasionally a point is retained after $p+1$ moves of the simplex. In this case either the experiment is in error (we never do perfect experiments) or the point really is a true optimum so the simplex has converged. There are various rules for how to deal with such cases, but it must be emphasized that simplex methods are designed for reasonably smooth and well behaved surfaces. We assume that there are no sudden discontinuities in the response surface. If there are, disasters can occur, and the simplex approach is probably not very useful. However, much of chemometrics assumes that natural processes are relatively well behaved and continuous processes, which is probably reasonably true in many practical situations. Elaborate stopping rules have been presented in the literature. These are certainly necessary to prevent computer programs collapsing and ensure that an answer is found no matter what the nature of the data. In the fixed sized simplex approach oscillation normally implies either that an optimum is reached or that we are searching the wrong part of the space, in which case a fresh start with different initial conditions and step-size may be advisable. Other rules involve prevention of boundary violations: for example it might be known that some of the factors have physical restrictions or that exceeding certain values is dangerous. These can all be built into the simplex algorithm but it must be emphasized that simplex approaches are best suited for relatively smooth and well behaved surfaces. The experimenter must not restrict himself to any one particular approach. If too many conditions and restrictions need to be placed on a fixed sized simplex optimization it is quite likely that an alternative approach is preferable.

A simplex optimization calculation for the data illustrated in Fig. 2.7 is given in Table 2.3. The simplex is repeated until no better response is found, as indicated by oscillation around a maximum, in this case. For convergence we stop when the response gets no better after three moves (there are, of course other possible convergence criteria which are outside the scope of this text). The movements are illustrated in Fig. 2.8.[1]

One problem not considered in the data of Table 2.3 is the problem of noise. We assume that every experiment is exactly reproducible in the simulation. This is not always the case, and this can cause some difficulties in simplex optimization, since repeating experiments under identical conditions will not

---

[1] In fact the response surface used to construct Fig. 2.7 and Table 2.3 was simulated using the equation $y = 9 - (x_1-2)^2 - 0.5(x_2-3)^2 - 0.2x_1x_2 + x_1$. It can be readily shown using simple calculus that the true optimum of this surface is given by $x = (2.245, 2.551)$ [$y=9.939$] which compares to the optimum found by simplex procedures of $x = (2.5, 2.598)$ [$y=9.870$]. The reason for this inconsistency is that the simplex procedure requires a fixed grid of possible triangles: different initial conditions and step-size would find different apparent optima. However, the answer in this case has been achieved fairly rapidly and is quite close to the true optimum.

always yield identical answers. This can cause severe oscillations if the optimum is very flat. Simplex methods do not build in assessments of experimental errors or replicates analysis.

The simplex procedure can be extended to several dimensions, or several factors. In this case, if the coordinates of the each of the $p+1$ points on the simplex are given by

$$\mathbf{x_q} \quad = \quad (x_{1q}, x_{2q}, \dots x_{pq}) \qquad (2.9)$$

where $\mathbf{x_q}$ is the vector of order p for the qth point on the simplex, it is easy to show that the new simplex is given by

$$\mathbf{x_{new}} \; = \; (2\sum_{q=1}^{q=p} x_{1q}/p \; - \; x_{10}, \; \dots, \; 2\sum_{q=1}^{q=p} x_{pq}/p \; - \; x_{p0}) \qquad (2.10)$$

where the $p+1$ previous points have again been ranked in order of optima, with 0 indicating the least favoured response and p the most favoured response, so that x_{iq} is the co-ordinate of $(p-q+1)$th best response for factor i. The new co-ordinate $\mathbf{x_{new}}$ is not yet ranked. Obviously a modification similar to Eq. (2.8) is used if the previous response was the worst one; in such a case $\mathbf{x_1}$ and $\mathbf{x_0}$ change around. For a one dimensional simplex it is easy to see that this reduces to

$$\mathbf{x_{new}} \quad = \quad \mathbf{x_2} + \mathbf{x_1} - \mathbf{x_0} \qquad (2.11)$$

which gives Eq. (2.7) for a two factor experiment.

There are many different *stopping rules* in the literature. Above we terminated the optimization when the response was clearly oscillating around a maximum. In practice experiments are not exactly reproducible, so other criteria are necessary. It is possible to determine whether the response has converged to a roughly constant value. Many further elaborations are possible, but out of the scope of this text. Although it is certainly possible to attempt to automate simplex approaches the best methods involve the experimenter judging for him / herself whether a true optimum has been found. If very many experiments need to be performed the simplex approach is probably inefficient and other methods such as those discussed in Section 2.4 might reach an answer quicker. Intuitive knowledge of the system and the aims of experimentation plays an important role in chemometrics methods.

Simplex approaches can be used to explore the optima of response surfaces and involve varying all the p factors at the same time and take account of interactions. However, there are still several limitations. The most obvious is that this approach is still critically dependent on the choice of initial conditions and step-size. Consider the one dimensional optimization problem, where we varied pH by only 1 unit at a time. This really was rather slow, particularly at the beginning of the optimization. It might seem more logical to change the step-size according to the progress of the optimization. In this way a more appropriate grid can be constructed since the result is not restricted to units of the original step-size. This approach is called the *modified simplex* approach.

There are, in fact, several different ways of applying the modified simplex procedures and the choice of method is, as usual, up to the experimenter. The difference between this and the fixed step-size procedure can be shown by modifying Eq. (2.10) to give

$$x_{new} = ((1+\alpha)\sum_{q=1}^{q=p}x_{1q}/p - \alpha x_{10}, \; ...., \; (1+\alpha)\sum_{q=1}^{q=p}x_{pq}/p - \alpha x_{p0}) \quad (2.12)$$

where α is a coefficient of expansion or contraction. In Eq. (2.10) this coefficient reduces to 1, since the size of the simplex is kept constant throughout the optimization. In a two dimensional modified simplex optimization, though, if the response setting $\alpha=1$ in Eq. (2.12) is found to be better than at x_2 (i.e. the response is getting better) it seems sensible to expand the simplex as we are obviously moving towards the optimum, and the faster we move the better. If this response is intermediate between the response at x_1 and x_2 it seems that we are homing in on an optimum gradually and have no real need to move faster or slower, so we retain $\alpha=1$. If the new response is worse than than at x_1 we probably are moving too fast and so contract the triangle. If it is even worse than x_0 we are probably moving in the wrong direction. For a two factor case it is most usual to use the following values of α.

(1) If the response at $\alpha=1$ is better than that at x_2 then test $\alpha=2$; if the response is still better than that at x_2, retain $\alpha=2$; otherwise retain $\alpha=1$.

(2) If the response at $\alpha=1$ is better than that at x_1 but worse than that at x_2 then $\alpha=1$.

(3) If the response at $\alpha=1$ is better than that at x_0 but worse than that at x_1 then $\alpha=+0.5$.

(4) If the response at $\alpha=1$ is worse than that at x_0 then $\alpha=-0.5$.

There is no need for a condition corresponding to Eq. (2.8) as we now rank the response and set a value of α according to this rank. The points x_{new}, x_1 and x_2 are retained in the next simplex calculation and so on.

The modified simplex procedure described above is probably best illustrated graphically (Fig. 2.9). There is, of course, no reason to restrict the value of α to 2 or ± 0.5, but $p+1$ values are required in advance of the optimization. For three or more factors there will need to be several different values of α; such cases are rarely reported in the literature, most chemical examples being restricted to two factor experiments. More than the four rules described above are required. In addition to the obvious advantages of the modified simplex over the normal simplex procedure in terms of speed and accurately locating an optimum, stopping rules can be better defined. A common criterion used is when the step-size has been reduced to 1% of the initial step-size.

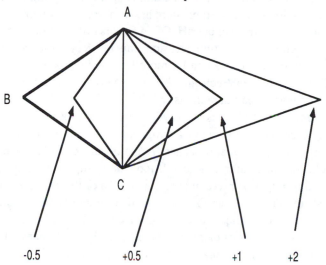

Fig. 2.9 - Modified simplex. The triangle ABC corresponds to the original three experiments. The new experimental points for α=+2, +1 and −0.5 as discussed in the text, are indicated.

In the modified simplex procedure we nevertheless fix the values of α in advance of experimentation and we do not use information about the shape of the surface. A more sophisticated approach is the *supermodified simplex* in which the value of α can be varied according to how rapidly the new response is being optimized. To use this approach it is normal to determine α by fitting a polynomial to the latest values of the response. Although this approach is theoretically yet more powerful, quite detailed knowledge of the response surface is now being built into the method, and other approaches including simultaneous methods described above might now become more appropriate. Simplex methods are probably most useful when the exact shape of response surfaces is of little interest and are not really designed for modelling the response. There are many elaborate rules for simplex optimization according to the problem and the sophistication of the approach desired. The chemometrician must never be conservative about methods. Simplex optimization is easy to perform. It can be achieved using pencil and paper or simple spreadsheets. There is no need to fit elaborate statistical models. It is quick and can sometimes reach an answer very quickly. However, if elaborations are being built into the system, it is useful to explore whether alternative methods should be used instead.

One difficulty with simplex procedures is that all factors need to be of roughly equivalent significance. If, for example, pH has very little influence over the rate of a reaction whereas temperature has a substantial influence, a great deal of time could be wasted changing pH. Often factors differ wildly in significance. Five or six factor simplex optimization can be very time consuming. An investigator might, in fact, have a lot more knowledge about the shape of this surface. For example, in Section 2.2 we discussed interaction effects, and suggested that not all the interactions will be significant. How can the simplex procedure be told this? There is no way in which this knowledge, often obvious to the chemist, can be exploited in the simplex approach. So some care should be taken when using simplex approaches. A major aim of experimentation is to reach conclusions in a minimum time. If too many simplex experiments need to be performed, it might be preferable to get an idea of the shape of the response surface instead. This can best be performed by quantitatively modelling the response surface using simultaneous designs as described in Section 2.4.

Despite these limitations there have been several spectacular demonstrations of the power of simplex optimization approaches, especially in the optimization of chromatographic separations and the tuning of scientific instruments.

A slightly different approach is called the method of *steepest ascent*.[1] In this case a local model of the response surface is determined. We start with one experiment and try to calculate the gradient on the surface at this point, to obtain a local *model* of the surface usually of the form

$$\hat{y} = b_0 + \sum_{i=1}^{i=p} b_i x_i \qquad (2.13)$$

This can be determined by using simple least squares regression analysis. The values of b_i determine the direction of maximum gradient, and the next experiment is chosen to be in this direction. In a simple two factorial experiment, the new value, $\mathbf{x_{new}}$, will be given by

$$\mathbf{x_{new}} \quad = (x_1 + \alpha b_1, x_2 + \alpha b_2) \qquad (2.14)$$

There are various similarities with simplex approaches. First, the method is critically dependent on the initial choice of starting point. Second, the adjustable parameter α determines how fast the ascent is. There has been much less literature on the use of this approach in chemistry, so elaborate rules for choosing α cannot be found, nor are there satisfactory rules for changing this value as the optimum is approached. Another weakness of the approach given by Eq. (2.13) is that the surface is modelled using a linear equation. Of course, higher order models can be established, but this is correspondingly slower and probably is equivalent to using a sledge-hammer to crack a nut. It is often better to employ simultaneous designs to produce higher order models of response

---

[1] Or steepest descent according to whether we are searching for a minimum or a maximum.

surfaces: the method of steepest ascent will only yield a local estimate of the gradient, and often a great deal of information is required to produce this estimate.

This approach does, though, have considerable advantages over the simplex method under some circumstances. First, if one factor has limited or no influence on the response this is easily taken care of, using the method of steepest ascent, whereas simplex methods assume each factor is equally significant. Second, higher order models can be used to take interaction effects into account. Again if the interaction between two factors is small this information will be available. A vital difference between this approach and simplex methods is that it is relatively easy to test whether the model is obeyed well or not. There are many approaches (described in other sections of this text) to assess how well a linear model is obeyed. If trouble is looming checks can be built into the system.

There are, though, few reports in the chemometric literature of the steepest ascent approach. This is probably because the attraction of simplex methods is their simplicity. More elaborate knowledge of the shape of response surfaces, such as that obtained by the method of steepest ascent, can certainly be put to use and will probably be more reliable and faster in finding an optimum, but the methods require greater computational sophistication and knowledge of the system that might not immediately seem relevant to the experimenter. How many synthetic chemists want or are interested in an equation for how the pH and temperature influences the yield of their reaction in a small region of response space, particularly since this equation is unlikely to be generally applicable? The chemometrician has a fearsome choice of approaches and methods. Scientific reasons are often confused with psychological and managerial reasons and it is impossible to disentangle the influence of these non-scientific factors on the choice of method, or, indeed define what really is the "optimum" result obtained from experimentation.

There are several other sequential approaches available to the chemometricians, most notably a method called evolutionary optimization (EVOP) but few sequential methods have been applied as frequently as the simplex procedure and the related modified and supermodified methods.

2.4 SIMULTANEOUS DESIGNS

Chemists tend by nature to prefer sequential designs to simultaneous approaches. Simultaneous designs generally involve performing a set number of experiments (typically about 20), recording the response from each of these experiments and then using a computer program such as a method for regression analysis to fit a model to the experimental data and then predict parameters such as the optimum or the value of interaction effects.

The simplex approach to optimization can be formulated in terms of simple and easily understood rules: chemists feel more secure with rules, as befits the psychology of what motivates most people to become chemists. Simultaneous designs involve investing a finite and foreseeable amount of effort before any results are forthcoming: it is possible to optimize a simplex in an indefinite number of moves, and experimenters can always hope that these moves will be few, if the initial guess is sufficiently good. In simplex approaches the optimum actually employed is observed: using simultaneous designs the optimum is often obtained from a curve fitting program. Many chemists are suspicious of "virtual" results. They like to see spectra and to look at compounds and until these objects are in their hands they rarely believe predictions, however solid the theory. So chemists like to see optima happening with their own eyes.

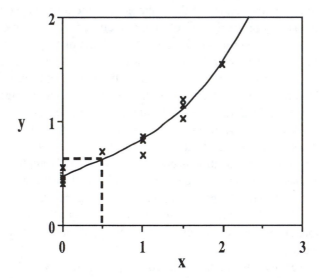

Fig. 2.10 - Accuracy of predicted point is greater than experimental point.

Yet simplex approaches can provide misleading results. There certainly is no guarantee that they will reach an optimum any faster than simultaneous methods. The simplex methods are best employed when there is some prior knowledge of a system, such as which factors are most significant. In simultaneous approaches good guesses can help achieve results faster, too, and a great deal of intuitive knowledge of the system can be usefully employed, such as the magnitude of interactions. When we followed the simplex in Table 2.3, we noted that we did not exactly find the true optimum because of the initial choice of the simplex.

Employing curve fitting and modelling, optima can be found more accurately as they do not need to be at the exact experimental point.

There is, however, a large cultural gap between the experimental scientist and the statistician which the chemometrician has to bridge. The chemist really wants to carry out an experiment before he believes it, no matter how good the theory is. This is deeply ingrained within classical experimentation in which the experiment, if properly carried out, is always more accurate than the theory, and, as discussed in Chapter 1, is the basis of traditional physical science. But the statistician will rely more on prediction. In Fig. 2.10 several experiments are illustrated with the resultant fitted curve. As we can see, the values of the intermediate points can be *predicted*. These predictions are probably more accurate than the experiments themselves, because there is a large experimental error. Therefore in many situations it is probably just as good to predict the response for a given set of factors as to actually measure it: if the experiment has been performed sufficiently well there will be information both about the analytical error and also about the genuine underlying trend, and this information will result in reliable predictions. The experimental process will be subject to a large error so any individual reading will be fairly suspect. In simplex optimization, described above, each individual response will also be subject to experimental error so, in fact we could go back to any combination of points **x** and obtain different values for the response. How bad is this error? Has any systematic information been collected to determine this? The answer is, of course, no. To the traditional chemist, schooled in an error free world (or a world in which he is unwilling to admit to the existence of errors) this approach of thinking is dangerous heresy, so much persuasion is often necessary.

Chemists do not only use simultaneous designs for optimization, but also when determining the value of quantitative parameters. In Section 2.2, we discussed the problem of linear calibration; the magnitude of a correlation coefficient between two parameters depends, in part, on how the experiment has been designed: how many unique points have been recorded, how many replicates have been observed? It depends on the size of experimental error (or analytical error), which, as we have seen, is often fairly significant. Yet again, though, traditional ways of working have proved a barrier to the use of experimental design in quantitative experiments. The most a chemist might normally do is use a curve fitting package and cite some statistic such as a correlation coefficient or χ^2 as evidence as to how well a model has been obeyed. Yet experimental errors, replicates information, the nature of the noise and the nature of the sampling is rarely considered. However, the method for interpretation and analysis of the data depends on the quality of the data and so how the data have been obtained. Analysis and design are inseparable. The better the quantitative experiments have been designed the more useful the resultant quantitative conclusions will be. Thus experimental design has a vital role to play in quantitative chemistry as well as in optimization studies.

There are several types of experimental design, but the first step is to code the factors and response, as in any design including, of course, sequential designs (see Section 2.2). Coding of the factors is especially crucial for simultaneous designs as the experiments determine the entire region of study in advance. If the optimum is far from the initial guess in simplex optimization this really does not matter. For sequential designs all responses of interest should lie within the region of the experiments: for example if the lowest coded pH is 4 and the highest is 8 in an experiment, there is little chance of reliably predicting an optimum at around pH 10. So intuitive ideas are put to great effect during the process of coding.

The next step is to consider the number of factors and the type of model to be investigated. Generalized models are given in Eq. (2.2). The problem is to determine the *order* of the model. For example a second order (quadratic) model for three factors is given by Eq. (2.4). From this equation it is possible to estimate the number of *coefficients* in the model, which for a second order three factorial model is 10, as can be verified from Eq. (2.4). This type of information is the key to effective simultaneous experimentation. In order to determine these parameters from the data at least 10 independent measurements need to be recorded. This should be fairly self-evident: it is impossible to fit a q-term model to less than q points. If 10 points are recorded the model will always exactly fit the data: equivalently it is always possible to exactly fit a straight line to two points. There is no information as to whether this model actually does provide a useful description of the data, so a few more experiments are useful, in the same way as it is useful to record three or four experiments in the case of linear calibration discussed in Section 2.1. But this does not provide information about replication or experimental error, so a few replicated experiments are useful also. The more replicates the more the information about the accuracy of the experiment so the better we can understand whether a poor fit to the data is actually due to a poor model or merely experimental error.

If we perform N experiments in total, of which N_R are replicates and there are Q parameters in the model (the number of coefficients in Eq. (2.2)), then there will be

$$\nu \qquad = \qquad N - N_R - Q \qquad\qquad (2.15)$$

experiments left to tell us how well the model has been fit. The parameter ν is generally referred to as the number of *degrees of freedom* used to assess how well the model is obeyed. For example, if four non-replicated experiments are used for a univariate linear experiment then $N = 4$, $Q = 2$ and so $\nu = 2$. Normally it is useful to have four or five degrees of freedom left to assess how well the model is fit. The resultant statistic is called the *lack-of-fit* statistic and will be discussed in greater detail in Section 2.5.

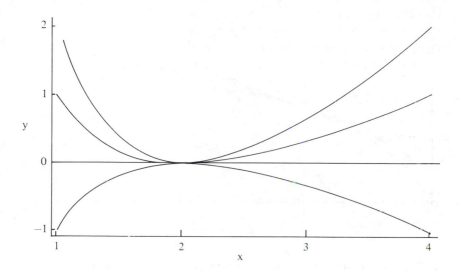

Fig. 2.11 - Graph of $x' = x' + b_{11}x^2$ for $b_{11} = -1, 0, 1, 2$ (from bottom to top). The parameter x is coded so that $x' = 2 \log (x/2) / \log(4)$, where all logarithms are to the base 10.

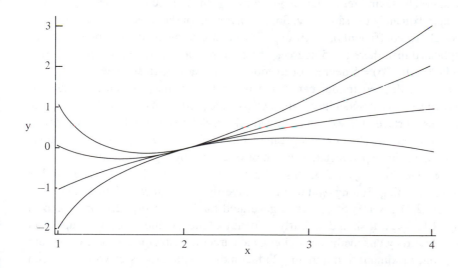

Fig. 2.12 - Graph of $x' = 2x' + b_{11}x^2$ for $b_{11} = -1, 0, 1, 2$ (from bottom to top). The parameter x is coded so that $x' = 2 \log (x/2) / \log(4)$, where all logarithms are to the base 10.

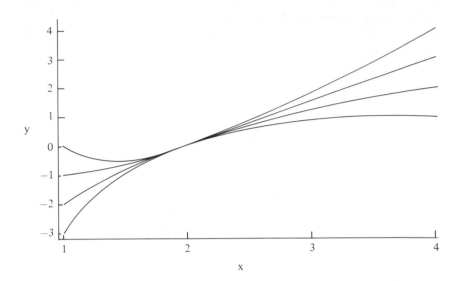

Fig. 2.13 - Graph of $x' = 3x' + b_{11}x^2$ for $b_{11} = -1, 0, 1, 2$ (from bottom to top). The parameter x is coded so that $x' = 2 \log (x/2) / \log(4)$, where all logarithms are to the base 10.

The number of degrees of freedom used to estimate the experimental error (assessed by taking replicates) is given by N_R. In an effective design N_R should be approximately the same as v. So, for example, for the three factorial quadratic model (10 coefficients), if 20 experiments are performed, five of which are replicated then there are 5 degrees of freedom to assess the experimental error and $(20 - 10 - 5) = 5$ degrees of freedom for the lack-of-fit statistic.

The desired model can guide us as to the most sensible number of experiments. The model itself will then provide predictions as to the shape of the response surface. For example, a quadratic model can be used to fit a variety of observed curves. The ratio b_i/b_{ii} can be regarded as a *shape* parameter whereas b_i is a magnitude parameter. This is best visualized by logarithmically coding the factors (see Eq. (2.6)), although similar reasoning is possible for linear and other forms of coding. Setting b_0 at 0, we can investigating changing these parameters. In Fig. 2.11 a series of curves is generated for $b_i=0$, in Fig. 2.12 $b_i=1$ and in Fig. 2.13 $b_i=2$. It should be fairly evident that most common curve shapes can be obtained by this method, providing the curves are either monotonic or contain only one maximum or minimum. In fact most natural processes will be of such type. More complex multioptima response curves are often quite unusual unless two separate processes are competing. Continuous variation of a factor such as pH or temperature is likely to result either in a single optimum or a continuous increase / decrease in a response. Obviously, multioptima shapes can be modelled using cubic terms or even sine waves, but these are rare in chemistry and normally we do not have sufficient specific knowledge of a system to assume such detailed models.

Some thought should be given to exponential and similar decay curves. Although it might be tempting to model responses using multiexponential functions, in practice these types of curve are very hard to fit. There is a detailed mathematical literature on the fitting of such models, but unless noise levels are low and the exponential coefficients are very different, it is hard to obtain reliable computational results. This type of modelling is referred to as *non-linear* modelling whereas models of the form of Eq. (2.2) are *multilinear* models. If an exponential trend is suspected it is probably computationally easier to take logarithms of the response and fit a (multi)linear model, than to analyse the raw data directly. This demonstrates the importance of coding both the response and the factors, to yield data that can be easily analysed computationally. Overlapping exponentials are obviously even harder to deal with. Some chemists may feel their experiment is inadequate if it cannot give an adequate physically precise answer to, for example, two overlapping exponential decay mechanisms: chemometrics should help convince the experimenter that there is little he or she can do about this. Possibly if the interpretation of a natural process is dependent on fitting a double exponential model to an observed parameter (e.g. the change in concentrations of chemicals over geological time has been used to interpret geological kinetic processes), the experiment is over-ambitious and a new experiment that does not require this type of modelling should be contemplated. Unfortunately the deep seated desire of many chemists for precise physical models can lead to potentially misleading and uninterpretable experimental data.

Once a sensible number of experiments, based on the desired model, is determined, it is necessary to choose a specific design. There are a number of designs to choose from.

Probably the simplest is a *factorial* design, as described below. As usual, one of the first stages is to *code* the variables. This coding may be logarithmic or linear, but the result of coding is that there are several *levels* of each factor. A level is a quantity; the experiment is designed so that each factor is observed only at several discrete levels.

The most comprehensive and straightforward factorial designs normally involve measuring the response for all possible combinations of factors at all levels. Therefore a P factor, k level design involves performing k^P experiments. So a three factor, three level design would involve 27 experiments. The conditions for any given design are normally expressed in tabular form. A two factor, three level design is given in Table 2.4 and involves nine experiments. The values of x_1 and x_2 are coded values of the factors determined by the method of coding chosen by the experimenter. These designs can also be illustrated in diagrammatic form. The two factor, three level design can be visualized by a square, each experiment being either on the four corners or on the four edges or in the centre (nine experiments in total) as is illustrated in Fig. 2.14. It can be seen that this design is adequate for a quadratic model which, for two factors, requires six terms, but inadequate for a cubic model unless some interaction terms are assumed to be negligible.

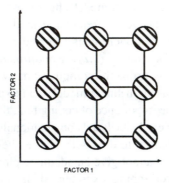

Fig. 2.14 -Two factor, three level design. Each circle represents one experiment.

Table 2.4 - Factorial design : two factor, three level

Experiment	x_1	x_2
1	+1	+1
2	+1	0
3	+1	−1
4	0	+1
5	0	0
6	0	−1
7	−1	+1
8	−1	0
9	−1	−1

There are two serious weaknesses of the factorial design. The first is that often a very large number of experiments is required to plan a full factorial design. For example, a four factor, four level design requires 256 experiments, which is clearly impracticable and far more experiments than necessary for a sensible model. As we will see below it is possible to reduce the number of experiments by employing partial factorial designs. A second difficulty is that there is no replicate information and so no estimate of the experimental error. This is overcome by performing replicate experiments often in the centre of the design (i.e. at $x = (0, 0..., 0)$).

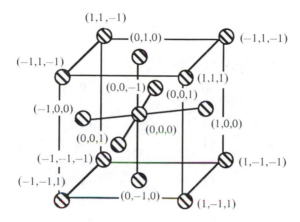

Fig. 2.15 - Face centred cube design (five replicates are taken in the centre).

However a more efficient approach is to perform less experiments over the experimental region. When there are only a few levels and a few factors, the factorial design probably provides sufficient measurements with which to estimate the model, but when the number of factors or levels is large there are far too many experiments, so some experiments can profitably be left out. A three factor, three level design is often represented by a cube in space. Each corner of the cube represents an experiment (eight experiments), the centre of each face (six experiments), the centre of each edge (12 experiments) and the centre of the cube also represent experimental points, making 27 experiments in total, as predicted. However, there are sufficient experiments if the 12 experiments in the centre of the edges are omitted, reducing the number of experiments to 15. If five replicates are included in the centre, this design is called a *face centred cube* (Table 2.5 and Fig. 2.15). For a p-factorial design of this type, there will be one central point, 2^p points on the corners of the hypercube, and $2p$ points on the faces of the hypercube plus any replicates. This design will be adequate to model multifactorial response surfaces.

Table 2.5 - Face centred cube design

Experiment	x_1	x_2	x_3
1	+1	0	0
2	−1	0	0
3	0	+1	0
4	0	−1	0
5	0	0	+1
6	0	0	−1
7	+1	+1	+1
8	+1	+1	−1
9	+1	−1	+1
10	+1	−1	−1
11	−1	+1	+1
12	−1	+1	−1
13	−1	−1	+1
14	−1	−1	−1
15	0	0	0
16	0	0	0
17	0	0	0
18	0	0	0
19	0	0	0
20	0	0	0

A further consideration is the level of the individual experiments (i.e. the values of each of the factors under which each experiment is performed). So far we have restricted these to discrete levels (−1, 0, +1), but there is no reason to do this. There are various statistical methods for determining the optimality of a design; these are very much linked with the modelling of experimental errors. Besides optimization, another aim of experimental design is to help guide the experimenter as to the most efficient and reliable way of constructing a quantitative model of a process, and if this model is constructed with equal confidence over the region of the design the experiment is probably fairly useful. If one part of the experimental region is modelled significantly better than another part then the design is rather asymmetrical. If the optimum, for example, happens to be in a poorly modelled part of the design it will be located less precisely than if in a better modelled region of the design. It is likely that the centre of the design will be best modelled, but the position of the outliers can critically influence the effectiveness of the experiment in the outlying regions. A statistically more satisfying design is the *central composite* design which is identical to a face centred cube, except that the points at (±1,...0) are changed to ($\pm4\sqrt{2^P}$,...0) where p is the number of factors. These points are at ±1.41 for a two factorial central composite design and ±1.68 for a three factorial central composite design: five replicates are taken in the middle again. So, for example, experiments 1 to 6 in Table 2.5 are changed so that experiment 1 is performed at $\mathbf{x}$ = (+1.68, 0, 0) rather than (+1, 0, 0) whereas experiments 7 to 20 are unchanged. The design can be altered still further. Indeed almost any arrangement of replicates and unique experimental points can be visualized, but it is probably best to use well established designs when first entering this field. It is worth mentioning *star* designs which are normally three level designs that contain just the central points and points at ($\pm a$, 0,...0), and so consist of the experiments on the edges of hypercubes (or geometrically related to these), and therefore consist of $2p+1$ points plus replicates in total. In the case of a three factorial design this results in an inadequate number of experiments to completely estimate the response parameters, unless certain interaction terms are assumed to be 0. Asymmetric designs are also possible, in which the positive and negative values of each factor are treated differently. However, if the chemometrician finds himself using elaborate asymmetric designs he will probably be best off looking at the method of coding used: it is always possible to code parameters in an asymmetric manner and so employ standard well established symmetric designs.

Another way of reducing the number of experiments based on a factorial design is by performing *fractional factorial* designs. It can verified that for a k level factorial design, involving k^P experiments, there are only sufficient degrees of freedom to fit terms which contain a maximum power of $x_i^{(k-1)}$, so a two level design can be used to only to estimate parameters of a maximum power of 1 in x_i. In other words, the model for a p factor, k level design will be

$$\hat{y} = b_0 + \sum_{i=1}^{i=p} b_i x_i + \sum_{i=1}^{i=p} b_{ii} x_i^2 + \dots + \sum_{i=1}^{i=p} b_{i\dots} x_i^k + \sum_{i=1}^{i=p-1} \sum_{j=p-i+1}^{j=p} b_{ij}$$

$$+ \sum_{i=1}^{i=p-1} \sum_{j=p-i+1}^{j=p} b_{iij} x_i^2 x_j + \dots + \sum_{i=1}^{i=p-1} \sum_{j=p-i+1}^{j=p} b_{ij} x_i^2 x_j^2 + \dots$$

$$+ \sum_{i=1}^{i=p-1} \sum_{j=p-i+1}^{j=p} b_{i\dots j} x_i^k x_j^k + \quad \dots\dots \tag{2.16}$$

This equation can be divided into one factor, two factor, three factor to p factor interactions, each of which contain terms up to order k in each factor. For a three factor, two level design the eight terms will be as follows:
(1) one intercept term,
(2) three linear terms in x_i,
(3) three cross-product terms in $x_i x_j$,
(4) one cross-product term in $x_i x_j x_k$.

Table 2.6 - Three factor, two level design

Experiment	x_1	x_2	x_3
1	+	+	+
2	+	+	−
3	+	−	+
4	+	−	−
5	−	+	+
6	−	+	−
7	−	−	+
8	−	−	−

However, as noted above, many of interaction terms are negligible in size. Let us assume that the four cross-product terms are negligible. Can we then reduce the number of experiments? Table 2.6 lists the full three factor, two level design. with + indicating a high level of each factor and − a low level of a factor (so, for example, if factor 1 was temperature which varied between 40 and 80ºC, factor 2 was pH between 4 and 8, and factor 3 catalyst concentration between 0.1 M and 0.5 M, experiment 4 would involve setting the conditions at 80ºC, pH 4 and a catalyst concentration of 0.1 M). Careful inspection of the table shows that each of the three factors has a unique sequence in the experiments: the sequence for factor x_1 is [+ + + + − − − −] whereas that for x_2 is [+ + − − + + − −]. Consider, now, the four cross-product terms. The *levels* of the cross-product terms are given by multiplying the levels for the appropriate x_i terms, so for experiment 1, the level of $x_1 x_2$ is +, so this term yields a sequence [+ + − − − − + +] using this shorthand. Thus all terms can be identified by a unique sequence, as can easily be verified. Consider now reducing the design to four experiments only. A symmetric way of selecting these experiments would be to pick two experiments in which each parameter is high and two in which each parameter is low. Table 2.7 lists a reduced (fractional factorial) experiment with the cross-product terms calculated. The column for $x_i x_j$ is exactly equivalent to the column for x_k: we say that $x_i x_j$ is *confounded* with x_k. The three factor interaction is, in fact, confounded with the intercept term. Thus we cannot distinguish between the effects for x_1 and $x_2 x_3$ using this design. If two and three factor interactions are thought to be negligible then this simplification has reduced the number of experiments required. In an equivalent way the number of experiments in a two level, five factor design can be reduced from 32 to 16 by confounding one factorial terms with four factorial terms, and two factorial terms with three factorial terms, so that, for example, $x_1 x_2$ would be confounded with $x_3 x_4 x_5$. This would assume that three and four factor interactions are fairly negligible.

Table 2.7 - Three factor, two level partial factorial design

	x_1	x_2	x_3	Cross-product terms			
				$x_1 x_2$	$x_1 x_3$	$x_2 x_3$	$x_1 x_2 x_3$
Experiment							
1	+	+	+	+	+	+	+
2	+	−	−	−	−	+	+
3	−	−	+	+	−	−	+
4	−	+	−	−	+	−	+

The rules for a constructing a half factorial design for p factors (two levels) can be summarized as follows.

(1) Construct a full factorial design for $p-1$ factors, which will involve 2^{p-1} experiments.

(2) Set the value of the pth parameter to be the product of the $p-1$ parameters. Thus x_p will be confounded with $x_1 x_2 ... x_{p-1}$ and so on.

There are, of course, other ways of constructing fractional factorial designs. It is possible to set the pth parameter to be minus the product of the other parameters; it is possible to produce asymmetric fractional factorial designs; the number of experiments can be reduced still further to quarter and even eighth factorial designs. Table 2.8 illustrates a five factor, two level, quarter factorial, design in which x_4 is confounded with $x_1 x_2 x_3$ and x_5 is confounded with $x_1 x_2$: there are, of course, many other possible quarter factorial designs, but it is important to think carefully about which cross-product terms should be confounded.

Table 2.8 - A possible five factor, two level, quarter factorial design

	x_1	x_2	x_3	x_4	x_5
Experiment					
1	+	+	+	+	+
2	+	+	−	−	+
3	+	−	+	−	−
4	−	+	+	−	−
5	+	−	−	+	−
6	−	+	−	+	−
7	−	−	+	+	+
8	−	−	−	−	+

When should partial factorial designs be used in chemistry? One of their weaknesses is that, if all terms are to be estimated, there are no degrees of freedom left to estimate the lack-of-fit to the model. Another weakness is that there is no replicate information. Finally the power of the individual terms is limited to 1 for the usual two level factorial design. In this way these designs do not provide the sort of quantitative information like designs such as the central composite or face centred cube design. But nor do simplex methods, and factorial designs do nevertheless provide some sort of model for the response even if we are unable to test how good this model is.

A fractional factorial design might, for example, be used to give quite an accurate picture of how HPLC chromatographic resolution varies with factors such as temperature, pH, solvent composition, flow rate and so on. Sequential methods will not provide this type of information and designs such as the central composite design are more oriented towards deducing quantitative parameters.

There are a large number of other formal designs cited in the literature. However, the main considerations must be the model, whether the terms in the model can be estimated, according to the number of degrees of freedom, and whether information on replicates is required. Factorial and fractional factorial designs tend to provide a general picture and are a good aid to experimenters who want to build a general model, whereas designs such as the central composite or face centred cube design (also called *response surface* designs) are useful for quantitative model building. A major difference between simultaneous and sequential designs is that it is necessary to analyse the data from a simultaneous design, normally *via* computational curve-fitting, subsequent to the experiments, whereas in designs such as simplex approaches, the answer (e.g. the response at an optimum) is immediately available experimentally. In this section we have only discussed how to set up a design – in the next section we discuss how to analyse the results of these experiments.

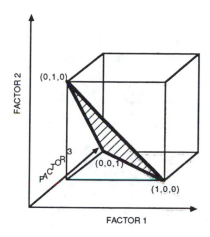

Fig. 2.16 - Mixture triangle.

Finally we discuss a type of design of particular interest to chemometricians, called a *mixture design*. Frequently chemists deal with proportions, for example a three solvent mixture. There are limits to the possible combination of factors. If, for example the three solvents are methanol, acetone and water, if there is $\alpha\%$ methanol then the proportion of acetone, $\beta\%$, must be $\leq (100-\alpha)\%$. The proportion of water will then automatically be $(100-\alpha-\beta)\%$. The three proportions are not independent. The composition of a p-component mixture can be represented by a simplex in $p-1$ dimensional space. For a three solvent mixture this is a triangle in two dimensional space (Fig. 2.16). A simplex design guides the experimenter as to what sensible proportion of mixtures to test in order to get an overall feel for the response surface. There is some confusion in terminology according to authors, but mixtures designs are normally of p-component *simplex centroid* designs: a three component design is illustrated in Fig. 2.17. Each "corner" of the triangle corresponds to an experiment consisting of 100% of one component. The centre corresponds to an equimolar mixture, i.e. 33.33% of each constituent. The centre of each edge corresponds to 50% of

two constituents. The design in Fig. 2.17 can be summarized in tabular form
(Table 2.9). A p-component design would consist of the points
(1) (1,0,..,0), (0,1,..,0) , , (0,0,..,1) [corners of the simplex],
(2) (0.5, 0.5,.., 0), (0.5, 0, 0.5.., 0), , (0, 0,.., 0.5, 0.5) [edges of the
    simplex],
(3) (0.333, 0.333, 0.333,.., 0), .., (0, 0,...., 0.333., 0.333, 0.333) [faces of the
    simplex],

(p) (1/p, 1/p, 1/p,.., 1/p) [centre of the simplex],
where the co-ordinates refer to the proportion of each component in the mixture.

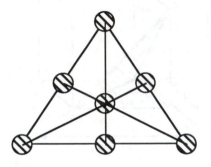

Fig. 2.17 - Simplex centroid design. Each circle corresponds to an experiment.
The triangle corresponds to the cross-section in Fig. 2.16.

Table 2.9 - Simplex centroid design corresponding to Fig. 2.17

Experiment	%A	%B	%C
1	100	0	0
2	0	100	0
3	0	0	100
4	50	50	0
5	50	0	50
6	0	50	50
7	33.33	33.33	33.33

These designs are efficient ways in which to scan a mixture space. The
resultant response, which could be the yield of a reaction or a chromatographic
response function, can then be modelled as for any other design taking into

account degrees of freedom, order of the terms in the model and so on. There are a variety of other mixture designs, but these have not as yet been exploited extensively in chemistry.

2.5 ANALYSIS OF RESPONSE DATA

The objective of much chemical experimental is to produce a quantitative model of a system. All the simultaneous designs described above may be employed to provide quantitative measurements of responses: these may, for example, be the rate of a reaction, the resolution of a spectroscopic peak etc. There are a number of statistical approaches for the analysis (quantitative modelling) of the resultant data.

The most usual approach to modelling is called *multilinear regression*. The responses are used to construct a model of the form of Eq. (2.2). The estimated value, $\hat{y}_n$, for experiment n, is related to the observed value, y_n, by the equation

$$y_n \quad = \quad \hat{y}_n - \varepsilon_n \tag{2.17}$$

where ε_n is the *error*. Normally this error is estimated using a least squares criterion so that the value of

$$S \quad = \quad \sum_{n=1}^{n=N} \varepsilon_n^2 \tag{2.18}$$

is minimized. It is important to note that it is normal to assume that there are no errors in the x axis or in the measured factor. Therefore if y is the yield of a reaction at different temperatures it is assumed that the temperature is always measured accurately. This assumption is not always valid, and more complicated methods for analysis are required under such circumstances: these are principal components based approaches and will be introduced in Section 4.2.

Any model can be used for $\hat{y}$, and the model is chosen according to the aims of the experiment. At this point it is important to appreciate that design and analysis are complementary to each other. Traditionally chemists analyse experiments (often by curve fitting) without taking the experimental design into account. However, we have seen from the discussion above, that the ability to fit a model of a given order depends on the number of degrees of freedom available. In the natural world there are almost certain to be interaction terms, and higher order polynomials with interactions can lead to a vast number of possible experiments. It is only necessary to compute the number of experiments required for a full two level factorial design: yet these experiments ignore terms in x^2, for example. In order to reduce the experimentation to manageable proportions we had to make various assumptions about which interactions are significant. Unless the problem is very simple it is almost inevitable that the chemist has implicitly made some assumptions about the size of interactions in his experiment, and the use of experimental design helps focus the mind sharply on these problems.

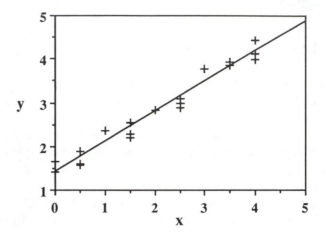

Fig. 2.18 - Linear model fitted to data with low experimental error.

A major aim of the interpretation of experimental data is to discover how well a model fits the data. There are various approaches to answering this question. The first is by calculating the root mean square error, or the root mean residuals. These are the root mean square difference between y (the observed value of y) and $\hat{y}$ (the predicted value of y)[1]. Fig. 2.18 illustrates a univariate linear model for which the mean square error is fairly small, whereas in Fig. 2.19 and Fig. 2.20 it is somewhat larger. It should be fairly obvious visually that the graph in Fig. 2.20 shows the possibility of a curvilinear trend. Fitting a quadratic model to Fig. 2.19 makes little difference to the mean square error, whereas it is dramatically reduced when applied to the data of Fig. 2.20, as illustrated in Figs 2.21 and 2.22. The raw data are tabulated for information (Table 2.10) and the reduction in errors given in Table 2.11. It is normal to divide S by the number of degrees of freedom used to assess the experiment which is $N–Q$, where N is the number of experiments and Q the number of coefficients in the model; 19 experiments were performed and for a quadratic model there will be three coefficients. The reason for this is explained below. Clearly the more terms in the model the lower the calculated error for a given set of experimental data under any circumstances, so employing a quadratic model reduces total error from 2.147 to 2.109 from Fig. 2.19 to 2.21. However this reduction in error is unlikely to signify a true quadratic trend, and, indeed is proportionately less than

[1]Normally S stands for the sum, s^2 for the mean sum of squares and s for the root mean square.

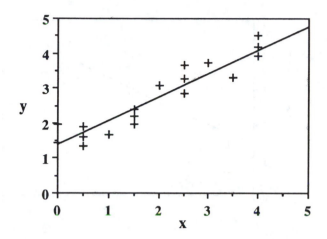

Fig. 2.19 - Linear model fitted to data with relatively high errors.

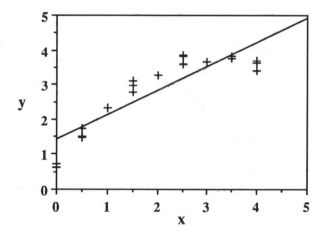

Fig. 2.20 - Linear model fitted to data that probably exhibits quadratic trends.

the change in number of degrees of freedom from 17 to 16. The reduction between Fig. 2.20 and 2.22 is much larger, from 4.755 to 0.249. The experimenter wants to determine whether this reduction in error is *significant* or is accounted for by the noise. Is the error within the limits of the experimental noise or not? Is there evidence for curvilinear trends?

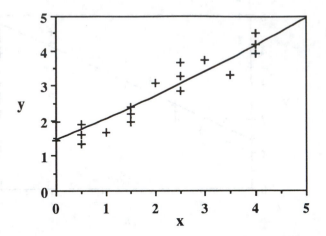

Fig. 2.21 - Quadratic model fitted to data of Fig. 2.19.

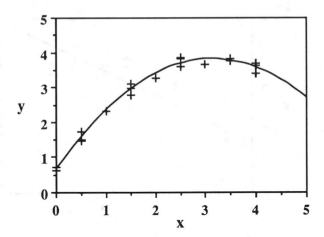

Fig. 2.22 - Quadratic model fitted to data of Fig. 2.20.

A statistical method for answering such questions is called *analysis of variance (ANOVA)*. The variance is defined as a mean square. For a parameter with p degrees of freedom it is given by

$$s^2 \quad = \quad \sum_{n=1}^{n=N} \xi_n^2 \, / \, p \qquad\qquad (2.19)$$

The parameter ξ_n could be the experimental error ε_n. The larger this error, the worse the model. But this error should not be treated as an absolute quantity. For example, if the mean square error for a linear fit to pH is 0.3 for a given experiment but for temperature is 10°C, is pH modelled worse than temperature? We need to *compare* this error with something.

Table 2.10 - Raw data for Figs 2.18 to 2.20

x	Fig. 2.18	Fig. 2.19	Fig. 2.20
0.0	1.411	1.433	0.629
0.0	1.644	1.945	0.707
0.5	1.892	1.589	1.471
0.5	1.599	1.906	1.503
0.5	1.582	1.338	1.721
1.0	2.357	1.654	2.321
1.5	2.278	2.391	2.788
1.5	2.544	2.182	3.114
1.5	2.201	1.972	2.968
2.0	2.824	3.063	3.282
2.5	2.992	3.270	3.837
2.5	2.880	2.854	3.867
2.5	3.093	3.646	3.600
3.0	3.762	3.715	3.651
3.5	3.926	3.311	3.764
3.5	3.847	3.288	3.839
4.0	4.437	4.199	3.685
4.0	3.982	4.516	3.400
4.0	4.116	3.934	3.639

At the beginning of this chapter we discussed the use of replicates when fitting a straight line. If the error due to *lack-of-fit,* which is the mean square error, is low compared to the replicate or analytical error, then the lack-of-fit is not very significant. The ratio between errors is called the *F-ratio*: the larger the ratio the more significant the variance. A well behaved model has a low F-ratio for the lack-of-fit statistic. Statisticians have constructed tables of F-values: these assume that errors are distributed in a Gaussian (or normal) manner. We will discuss these distribution curves in more detail in other chapters, but if errors for each experiment and at each sampling point are independent this is often the case.

It is possible to look at the *significance* of an F-ratio: the ratio will be the ratio of two variance or error terms of v_1 and v_2 degrees of freedom respectively, and $F_{v1,v2}$ can be looked up in standard tables of statistics. Normally these tables cite significance levels. The most popular significance levels are
(1) 95% (often denoted by *)
(2) 99% (**)
(3) 99.9% (***).

Table 2.11 - Mean square errors for data in Figs 2.18 to 2.22

Fig.	Type of model	Best fit equation	S	s^2	s
2.18	Linear	$y=1.410+0.690x$	0.579	0.034	0.184
2.19	Linear	$y=1.378+0.676x$	2.147	0.126	0.355
2.20	Linear	$y=1.415+0.699x$	4.755	0.280	0.529
2.21	Quadratic	$y=1.448+0.556x+0.029x^2$	2.109	0.131	0.362
2.22	Quadratic	$y=0.649+2.017x-0.321x^2$	0.249	0.016	0.124

Note: S is a total sum of squares, whereas s is the root mean square of the appropriate sum.

Significance levels are usually quoted after each cited value of the F-ratio. It is important to remember that the significance of this ratio depends on the number of degrees of freedom, in the same way that the significance of a correlation coefficient is dependent on the number of degrees of freedom.

So we have a more objective way to assess how well a model is obeyed, and this approach can take into account how the experiment has been designed. It is common to break the errors down systematically according to the terms in Eq. (2.15). The degrees of freedom for each of the sum of squares are summarized in Table 2.12. So, for a three factorial central composite design (20 experiments) with a model of the form of Eq. (2.4), there are 10 parameters in the model, thus, 10 degrees of freedom to estimate these parameters, five replicates and, therefore five degrees of freedom left to assess the lack-of-fit to the model since $10+5+5=20$. Normally the mean (b_0) is of little interest to the model, and is subtracted from the errors, so, in practice there are only 19 degrees of freedom in total, nine of which are used to estimate the remaining coefficients. Therefore an F-ratio with a denominator and numerator of five can be used to assess the significance of the lack-of-fit with respect to the replicate error. It is also possible to assess the significance of each of the nine parameters: each term reduces the overall error: the amount each term reduces the error by can be calculated, simply by removing the term from the model. The more this term reduces the error in estimating the model the more significant the term is. Clearly the significance of each term in the model can be estimated by using an F-test as well.

Table 2.12 - Degrees of freedom for an experiment

Parameter	Degrees of freedom
Total experiments	N
Parameters in model	Q
Replicates	N_R
Lack-of-fit	$N-N_R-Q$

These results are normally presented in tabular form (an ANOVA table). It is important to remember, though, how widespread the uses of ANOVA are: the method can be used for assessing how well a complex response surface is modelled or merely the fit to a straight line. It is far more flexible than the use of correlation coefficients which depend on degrees of freedom and also better than quoting absolute values of mean square errors, which are of little use if there is insufficient information about the analytical (experimental) error as assessed by replicate experiments.

The use of ANOVA is best illustrated by a simple example. Consider the data employed to produce Figs 2.20 and 2.22. These data are the results of 19 experiments, 10 of which are replicates. No effort has been made to optimize the design in this case, and there are probably more experimental points than are really necessary, but these data are typical of much laboratory experimental data. The data (x and y) are tabulated in Table 2.13, together with the predicted linear model ($\hat{y}_l$) and the predicted quadratic model ($\hat{y}_q$). In addition another parameter obtained from the replicates analysis is tabulated, which we will call y_r. This parameter is the *mean* value of y for any given value of x. We will denote $y_{r,n}$ as the value of y_r for experiment n; similar notation applies to the other parameters. If an experiment is not replicated, then clearly $y_{r,n} = y_n$; the root mean square

$$S_{\text{rep}} \quad = \quad \sqrt{\Sigma (y_n - y_{r,n})^2 / N_R} \qquad (2.20)$$

is the mean analytical error: clearly the greater this error the less reproducible the experiment (this does not imply that the experiment is poorly designed − this type of information is obtained from the experiment). The value of N_R is 10 in this case as there are 10 replicates, as can be seen by examination of Table 2.13. This value is *independent* of the model (linear or quadratic) chosen. However the root mean square error

$$s_{\text{lof}} \quad = \quad \sqrt{\Sigma \ (\hat{y}_n - y_{r,n})^2 \ / \ (N - Q - N_R)} \qquad (2.21)$$

may be used to assess the lack-of-fit to the model. This is the difference between the predicted value of y and the value of y averaged for replicates (so experimental / analytical variability is taken into account: the overall experimental results will depend *both* on how the system actually obeys the model *and* on the

Table 2.13 - Data for Figs 2.20 and 2.22 and Table 2.14

n	x_n	y_n	$y_{r,n}$	$\hat{y}_{l,n}$	$\hat{y}_{q,n}$
1	0	0.629	0.668	1.415	0.649
2	0	0.707	0.668	1.415	0.649
3	0.5	1.471	1.565	1.764	1.577
4	0.5	1.503	1.565	1.764	1.577
5	0.5	1.721	1.565	1.764	1.577
6	1.0	2.321	2.321	2.114	2.345
7	1.5	2.788	2.957	2.463	2.952
8	1.5	3.114	2.957	2.463	2.952
9	1.5	2.968	2.957	2.463	2.952
10	2.0	3.282	3.282	2.812	3.399
11	2.5	3.837	3.768	3.162	3.685
12	2.5	3.867	3.768	3.162	3.685
13	2.5	3.600	3.768	3.162	3.685
14	3.0	3.651	3.651	3.511	3.811
15	3.5	3.764	3.801	3.861	3.776
16	3.5	3.839	3.801	3.861	3.776
17	4.0	3.685	3.575	4.210	3.581
18	4.0	3.400	3.575	4.210	3.581
19	4.0	3.639	3.575	4.210	3.581

Note : The experiment number is given by n, the values of the factor for each experiment by x_n and the response by y_n. The value of the response averaged for replicates (analytical error) is $y_{r,n}$. The estimated value of the response for a linear model is given by $\hat{y}_{l,n}$ and a quadratic model by $\hat{y}_{q,n}$.

analytical / experimental technique). Clearly the greater this number the worse the fit to the model, but the number of terms in the model (three for the quadratic and two for the linear model) and the number of replicates, must also be taken into account. The remaining degrees of freedom are, therefore, six for the quadratic model and seven for the linear model.

Table 2.14 - ANOVA for data in Table 2.13

LINEAR MODEL

Parameter	Sum of squares	D.f.	Mean ss.	Mean square ratio
Predicted model	17.142	1		
Total error	4.755	17		
Lack-of-fit	4.569	7	0.623	35.09***
Exptl. error	0.186	10	0.0186	

QUADRATIC MODEL

Parameter	Sum of squares	D.f.	Mean ss.	Mean square ratio
Predicted model	21.462	2		
Total error	0.249	16		
Lack-of-fit	0.063	6	0.01005	0.54
Exptl. error	0.186	10	0.0186	

Notes
1. The predicted model is normally adjusted for the mean or b_0 by subtracting this from all values of y prior to analysis. Hence one degree of freedom is lost.
2. The lack-of-fit parameter is compared to the experimental error. The * corresponds to confidence levels as indicated in the text.

Sums of squares should be additive. Another parameter, calculated in Table 2.11, is the root mean square of the total error. This error is given by

$$ s = \sqrt{\Sigma(y_n - \hat{y}_n)^2/(N-Q)} \qquad (2.22) $$

which is identical to the relevant error in Table 2.11. But, $y_n - \hat{y}_n = y_n - y_r + y_r - \hat{y}_n$, so[1]

$$ S = S_{rep} + S_{lof} \qquad (2.23) $$

A number of other ratios and sums of squares can be calculated, but comparing sums of squares or *variances* is the basis of ANOVA.

These sums of squares are tabulated in Table 2.14, for both linear and quadratic models. Normally the intercept parameter (b_0) (mean) is removed from the analysis, so one degree of freedom is lost. The F-ratio of the root mean

[1]Remember that in this Section, S stands for the sum of squares.

square lack-of-fit to the root mean square experimental (replicate) error is then computed for both models. Standard statistical tables provide F distributions. Normally the critical values of F are tabulated, so levels of significance can be estimated, but many standard computer packages are available that provide probabilities for any given value of F and any number of degrees of freedom. In the case of the lack-of-fit statistic for the linear model, the value of $F_{7,10}$ is looked up, since a variance corresponding to seven degrees of freedom (lack-of-fit) is being compared with a variance of 10 degrees of freedom (experimental error). Normally far fewer replicates are required for a good estimate of experimental error, and well designed experiments often involve approximately equal numbers of degrees of freedom for estimating the lack-of-fit statistic and the experimental error. It is easy to verify that the lack-of-fit and experimental error add up to the total error. It is evident, from Table 2.14, that the lack-of-fit for the linear model is highly significant, but not for the quadratic model, suggesting that a quadratic model is a good model.

Of course many other ratios could be employed in ANOVA, indeed there is a large literature on this subject. For example, it is not only possible to calculate the lack-of-fit statistic, but also the goodness-of-fit for each individual coefficient in the model. This corresponds to how much each coefficient contributes to reducing the overall total error. Sometimes apparently contradictory results arise in chemometrics: there may be a significant lack-of-fit to the overall model but also a significant goodness-of-fit for individual parameters. There are many possible explanations for this. First, the model may not involve enough terms: extra terms might reduce the total experimental error and so the F-ratio. However, it is important to realize that the number of parameters in a model is, to a certain extent, limited by the experimental design. A three factorial central composite design (discussed above in Section 2.4) can only be used to estimate a maximum of 14 coefficients, and under such circumstances there are no degrees of freedom left to assess the lack-of-fit parameter. So design and analysis are intricately intertwined. A second and interesting aspect of chemometrics is that analytical (experimental) error is often relatively low compared with many other areas of science such as biometrics and psychometrics. This was discussed in Chapter 1. The ANOVA example illustrated above involved comparing mean sums of squares with the mean squared (replicate) analytical error. If this error is low due to relatively good experimental reproducibility, the entire analysis will be affected and several ratios will appear significant. In such circumstances it might be useful to test other ratios. For example, the significance of individual coefficients could be compared with the total error (lack-of-fit + analytical error); endless possibilities exist.

There are a large number of standard software packages available for calculating ANOVA tables and degrees of significance (e.g. SAS, MINITAB, GENSTAT) so it is unlikely that the reader will need to compute these parameters by hand. However, it is important to be able to interpret ANOVA tables and a simple example such as the one above should help this understanding.

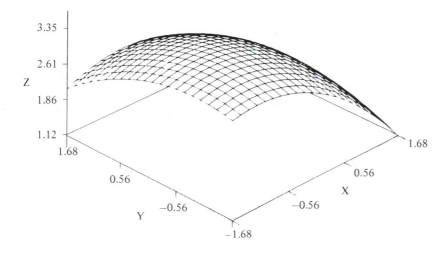

Fig. 2.23 - A simple response surface.

ANOVA has wide potential use in chemometrics. It can be used to compare sources of errors. For example repeated GC calibrations may be performed by several analysts on several instruments using various types of software. What is the major source of variability? Is it best to use only one analyst? Is the instrumental variability greater than operator variability? This type of problem will be discussed in greater detail in Chapter 4, including various sophistications such as cross-classification and hierarchical ANOVA. Finally sometimes the response itself is multivariate. Consider the case of the use of experimental design to explore a chromatographic response. There may be 20 peaks in the chromatogram. It would be possible to model the change in width or height of each peak separately, but, in practice, there are a few principal factors that influence these changes throughout the chromatogram, and these trends can be estimated using multivariate analysis as described in later chapters.

Once a reasonable model, often in the form of a polynomial of such as Eq. (2.2), is established, it is vital to interpret this model. There is little point designing and analysing complex experiments unless something *useful* arises from the experimentation. Probably the simplest approach is by displaying Eq.

(2.2) in the form of a graph of change in response as the value of each factor changes. For a single factor experiment this is merely a line. For two or more factors we obtain a multidimensional *response surface*. A two factor response surface is illustrated in Fig. 2.23. Often these surfaces are displayed as contour

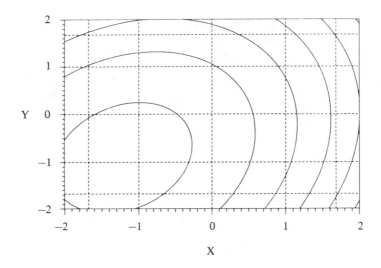

Fig. 2.24 - Contour representation of data from Fig. 2.23.

plots. The advantage of this latter method of display is that it is easier to "pinpoint" optima, but care should be taken when selecting contour levels; a contour plot of Fig. 2.23 is illustrated in Fig. 2.24. Some chemometricians use colour for visualizing the intensity of the response; for example, red could correspond to a high response and blue to a low response, with intermediate colours and shades corresponding to intermediate responses. Grey shades can also be used: the response could be coded as a number between 0 and 255, with 255 corresponding to black and 0 to white. It is vital to recognize the potential impact of facile computer graphics in this area. Much of classical statistics can, in fact, be regarded as geometry. Statistical parameters are translated into geometric distances: this aspect is particularly important in multivariate analysis, as will be discussed in subsequent chapters. However major statistical packages developed in the 1970s and 1980s, often involving many hundreds of manyears, were oriented towards batch processing on mainframe computers, and not readily available microprocessor based graphics. Most traditional statistical texts were produced prior to the ready availability of computer graphics, when plotting complex multidimensional surfaces was not an easy task. So the most a statistician might do is plot a contour diagram of a response surface. The rest of the information would be displayed as tables of numbers. Graphics represents a potential advance in conceptual thinking: for example chemists represent molecular structures by line diagrams rather than tables of electron densities – this approach enables the chemist to think more clearly about chemical reactions; recent computational advances in molecular graphics been employed as major aids in synthesis planning and understanding structure activity relationships of complex molecules. Thus the ability to easily visualize response surfaces as complex three dimensional structures in real-time is more than just a cosmetic advance, although many traditional statisticians might not acknowledge this.

It is important to recognize that reconstructed response surfaces are not directly observed, but best-fit models to the data. When more than two factors are involved this surface becomes a four or more dimensional graph, which, clearly, we cannot readily visualize. The normal approach is to take two or three dimensional projections in the multifactorial hyperspace, so for a three factorial design, there will be three possible three dimensional projections, corresponding to the three possible combinations of any two factors. However, there are several possible ways of taking three projections. Consider a three factor experiment. A response surface can be constructed of the change in response as x_1 and x_2 are varied, but what of the third factor, x_3? The value of this factor should be fixed at a given level. Normally parameters are *coded* as discussed above. The models given by Eq. (2.2) simplify considerably if x_3 is set to 0; a coded value of 0 does not, though, correspond to absence of the factor, but an "intermediate" value of this factor. If there are interactions between factors this will influence the shape of the resultant response surface. Sometimes an alternative approach is to plot the response surface at the optimum value of the third parameter (x_3). Under such circumstances, if the two factorial response surface exhibits an optimum also, this will be the same for any two combinations of factors providing all three two factor response surfaces are displayed at the optimum value of the third factor. Finally it is often desirable to display response surfaces in the absence of the third factor, but whether this is possible or meaningful will depend on the nature of the coding. Sometimes this would involve setting the third factor to a coded value of $-\infty$ in the model (which is based on coded values of the factors), which clearly is impracticable. Again the vital importance of considering the design prior to experimentation according to the type of answer required must be emphasized.

It is impossible to separate statistical from graphical software. Some of the more specialized statistical packages such as GENSTAT and GLIM are written by and for small groups of numerical statisticians and, although they provide excellent facilities for advanced matrix algebra, they cannot keep pace with modern graphical hardware and software. Many packages run on large computers, or are written in system independent languages such as C. The consideration of which software to use is a major one prior to extensive analysis of experiments. Software must have the ability to fit fairly complex surfaces in a reasonable time but at the same time substantial graphical output is necessary. Chemometrics is about helping the experimentalist reach conclusions in a short time. There is little point developing elaborate subroutines in low level languages, since this is expensive on time and unlikely to lead to novel chemometrics. Possibly the best approach is to use dedicated workstations and well supported packages such as S or SAS. Another possibility is to import data from one package and computer to another: for example the results of large statistical calculations could be transferred to a microcomputer for flexible graphical manipulations. Often people trained in one approach strongly resist change, so care must be taken when evaluating the best (most efficient) method. SAS probably is the most comprehensive combined statistical and graphical package available at present but is somewhat limited in interactive facilities.

    A final aspect of the analysis of response surfaces is the visualization of *confidence limits*. These tell us how good a model is, given the observed errors. It is possible to fit an equation to any data, but it is constructive to see how closely the model fits the data. A 95% confidence interval means that we expect 95% of responses to be within this region: clearly the closer the confidence intervals are to the model, the more confidence we have in the model. Most confidence interval calculations assume that the errors are given by a *Gaussian* or normal distribution (see Chapter 3 for an introduction to such distributions). There are three types of confidence interval.

(1) *Individual predicted*. These confidence intervals are for the model taking into account analytical (experimental) error as assessed by replicates analysis. In other words if there is a high replicate error, this will increase the width of the confidence bands.

(2) *Mean predicted*. These are the confidence intervals for the observed response after replicates have been averaged out, i.e. for $y_{r,n}$ rather than y_n in the notation above.

(3) *Working-Hotelling*. These are for the response surface as a whole; the Working-Hotelling limits do not apply to any individual point, but to the overall reconstructed line or surface.

The 95% individual predicted and mean predicted confidence intervals for a straight line are illustrated in Fig. 2.25. Obviously these intervals can be set at any percentage level.

    It is, of course, possible to calculate these intervals for any response surface in any number of dimensions, and flexible multidimensional graphics can be

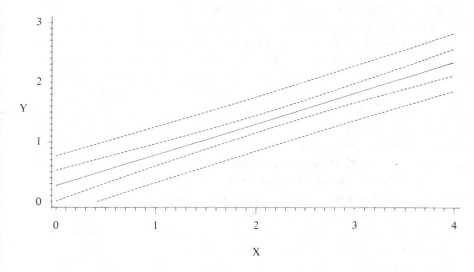

Fig. 2.25 - Confidence limits for a straight line : 95% individual predicted (outer limits) and mean predicted.

employed to visualize these intervals. It can be shown that the shape of these intervals depends solely on the design and number of parameters in the model (see bibliography for more details). For a Q–coefficient model, the various terms can be expressed as a vector. For example, for a one factor quadratic model in x_1 this vector will be given by

$$\mathbf{x} \quad = \quad (1 \quad x_1 \quad x_1^2) \tag{2.24}$$

and consists merely of the values by which the Q coefficients are multiplied for any given point on the surface. So for a three factor quadratic model of the form of Eq. (2.4) this vector becomes

$$\mathbf{x} = (1 \quad x_1 \quad x_1^2 \quad x_2 \quad x_2^2 \quad x_3 \quad x_3^2 \quad x_1 x_2 \quad x_1 x_3 \quad x_2 x_3) \tag{2.25}$$

This vector can, therefore, be calculated for any combination of values of x_i and is dependent on the model. The design may be represented by a *design matrix*. Each row of this matrix corresponds to the vector $\mathbf{x}$ for each given experimental point, and there are N rows in this matrix, where N is the number of experiments. Therefore this matrix will consist of N rows and Q columns: we call this the $\mathbf{X}$ matrix. This matrix is dependent on the design (the values of x_1, x_2, etc. for each of the N rows) and the model (the Q columns) and is the key to statistical analysis of experimental designs.

It is possible to define a value,[1] h, that can be calculated for any point x, given by

$$h \qquad = \qquad x (X'X)^{-1} x'$$

(2.26)

It can be shown that h is a scalar or single number rather than a matrix or vector. The value of h is given by the design and the number of parameters in the model. This is equivalent to a statistical parameter called *leverage* which can be calculated for any experimental point. Eq. (2.26) extends leverage to any point whether experimentally observed or not. This latter facility is a very important extension to the concept of leverage. There is little point running experiments if the only places where it is possible to predict the response are the actual experimental points. For example supposing we recorded the yield of a reaction at pH 5 and pH 6. Can the yield at pH 5.5 be predicted even though this was not an experimental point? What is the *confidence* in this prediction?

The confidence intervals are related to leverage. The lower the leverage, the higher the confidence in the predicted point, and the narrower the confidence intervals. This information is important to the experimenter. For example for many designs, most experimentation is in the centre of the design. Yet there may be considerable chemical interest in behaviour all over the experimental region. Is there enough information to model this behaviour well? In Fig. 2.25, the confidence bands diverge away from the centre of the regression line. The shape of these confidence bands is related to the design. This is yet another crucial link between design and analysis. This divergence can be understood by the link between confidence bands and leverage.

The value of the confidence intervals for the mean predicted is given by

$$y_{\pm} \qquad = \qquad s \sqrt{(F_{1,N-P}\, h)}$$

(2.27)

for the individual predicted (taking into account replicate error), it is given by,

$$y_{\pm} \qquad = \qquad s \sqrt{(F_{1,N-P}\, (1+h)\,)}$$

(2.28)

and for the Working-Hotelling limits it is given by

$$y_{\pm} \qquad = \qquad s \sqrt{(P\, F_{P,N-P}\, h\,)}$$

(2.29)

The value of the F-statistic can be set at any level, but 95% levels are normally employed. It is evident that these confidence bands depend on two things. First, s, is the overall root mean square experimental error as discussed above and is dependent on the results of experimentation. It determines the *magnitude* of the confidence limits. Clearly, the larger the experimental error, the broader the confidence intervals. Second, the shape is dependent only on the design and parameters in the model i.e. the value of h and the F-statistic and can be calculated prior to experimentation. Clearly the larger this is the less the confidence in the model.

---

[1] The *transpose* of a matrix or vector is denoted by ' and the inverse by $^{-1}$ as defined in the appendix.

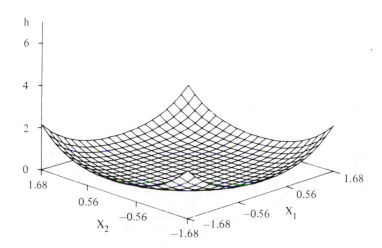

Fig. 2.26 - Graph of h for a central composite design.

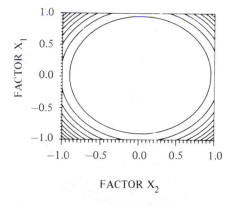

Fig. 2.27 - Contour representation of Fig. 2.26.

The value of h can be displayed graphically and provides a good indication of the effectiveness of the design. Fig. 2.26 is a three dimensional graph of leverage for a three factor central composite design (Section 2.4); the third factor

(x_3) is set at a coded value of 0. The corresponding contour plot is displayed in Fig. 2.27. It is evident that h is least in the centre but sharply increases on the

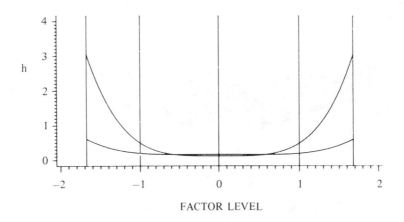

FACTOR LEVEL

Fig. 2.28 - Graph of h for face centred cube (upper right and upper left) and central composite designs for one factor where the other two factors are held at coded values of 0.

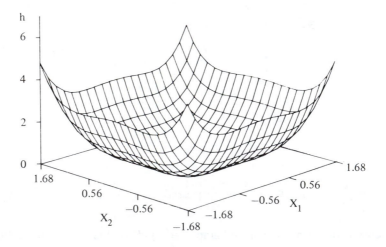

Fig. 2.29 - Graph of h for face centred cube design (two factor surface).

outside of the design. Thus there is a strong difference in confidence in the predictions over the region of experimentation.

What happens if the design is modified? As discussed above, the position of the outliers can be changed; if they are changed from ±1.68 to ±1.0 (see experiments 1 to 6 of Table 2.5) the design is called a face centred cube design. This dramatically changes the graph of h as illustrated in Fig. 2.28 for a one factor cross section (x_2 and x_3 are set at coded value of 0), and in Fig. 2.29 for two factors. It is possible to see the influence of design and model on predictions by constructing graphs of h. This demonstrates the importance of readily available graphics. Factors such as the number and position of replicates, the position of outliers, the model, the symmetry of the design and so on can be analysed in detail by this means.

2.6 THE INTEGRATION BETWEEN DESIGN AND ANALYSIS

Often experimental design is regarded as a separate part of chemometrics. In fact, as this chapter has shown, design and analysis go hand in hand.

Obvious uses of experimental design in chemistry include optimization and modelling. Optimization is important in many areas of chemometrics, for example when tuning scientific instruments: the width of a peak can be optimized according to instrumental parameters. The sharper and so better resolved the peak the more accurate the quantitative information. There is normally a trade-off between design and analysis under such circumstances. Approaches such as deconvolution (described in Chapter 5) can also be used to provide quantitative parameters from instrumental signals. The chemometrician must weigh carefully the pros and cons of spending time on experimental optimization and on data analysis under such circumstances. A great deal depends on the questions being asked. Quantitative modelling is another important area of chemometrics. In this chapter we have concentrated on *multilinear models* of the form of Eq. (2.2). However many of the principles of experimental design can be extended to all areas of multivariate chemical modelling and calibration. In Chapter 6 we will discuss partial least squares regression and principal components regression. These are ways of fitting models to complicated functions of a response and much used in chemistry. The problems of correlation coefficients, arrangement of experiments and replication apply equally well in such cases, although less theoretical work has been performed in this area. Possibly one difficulty with the evolution of chemometrics is that different aspects of the subject have been placed in "boxes", so there are specialists in multivariate analysis and in experimental design and most of these have been different people. The future of the subject will probably depend to a great extent on integration of all aspects of chemometrics. It is therefore important that chemometricians whose principal interest in the subject comes from analysis of data consider carefully the design of the experiment and use chemometrics as a tool for experimentation as well as analysis. Equivalently chemometricians interested largely in experimental design should consider whether the analysis of the experiments could be improved by using multivariate and other approaches. For example, much work has been performed on chromatographic optimization. Yet normally a single CRF (chromatographic response function) is calculated which is a univariate parameter, normally a function of widths or heights of peaks across a spectrum; obviously more information would be available if a multivariate parameter were analysed, such as a principal component. It is likely that such methods for *multivariate response surface methodology* will be developed and employed more in the future.

REFERENCES

General
S.N.Deming and S.L.Morgan, *Experimental Design : a Chemometric Approach*, Elsevier, Amsterdam, 1987
O.L.Davies (Editor), *The Design and Analysis of Industrial Experiments*, Oliver and Boyd, London, 1984
N.L.Johnson and F.C.Leone, *Statistics and Experimental Design in Engineering and Physical Sciences*, Wiley, New York, 1964
G.E.P.Box, W.G.Hunter and J.S.Hunter, *Statistics for Experimenters*, Wiley, New York, 1978

Section 2.1
M.A.Allus and R.G.Brereton, *Int. J. Env. Anal. Chem.*, **38**, 279 (1990)

Section 2.3
K.W.C.Burton and G.Nickless, *Chemometrics Int. Lab. Systems*, **2**, 135 (1987)
W.Spendley, G.R.Hext and F.R.Himsworth, *Technometrics*, **4**, 441 (1962)
D.E.Long, *Anal. Chim. Acta*, **46**, 193 (1969)
G.E.P.Box and N.R.Draper, *Appl. Stat.*, **6**, 81 (1957)

Section 2.4
E.Morgan, K.W.Burton and P.Church, *Chemometrics Int. Lab. Systems*, **5**, 283 (1989)
M.A.Allus, R.G.Brereton and G.Nickless, *Chemometrics Int. Lab. Systems*, **3**, 215 (1988)
J.A.Cornell, *Experiments with Mixtures: Designs, Models and Analysis*, Wiley, New York, 1981

Section 2.5
N.Draper and H.Smith, *Applied Regression Analysis*, Wiley, New York, 1981
R.H.Myers, *Response Surface Methodology*, Allyn and Bacon, Boston, MA, 1970
M.A.Allus, R.G.Brereton and G.Nickless, *Chemometrics Int. Lab. Systems*, **6**, 65 (1989)
H. Sheffé, *The Analysis of Variance*, Wiley, New York, 1959
H.Working and H.Hotelling, *J. Am. Stat. Assoc. Supplement (Proceedings)*, **24**, 73 (1929)
L.Ståhle and S.Wold, *Chemometrics Int. Lab. Systems*, **6**, 259 (1989)

3

Sampling sequential series

3.1 TIME SERIES IN CHEMISTRY

In Chapter 2 we discussed the problems of sampling over an experimental region, which should be clearly considered prior to acquisition of data. The experimental region was rather an artificial mathematical concept. For example if one factor in an experiment is pH, the experimenter arbitrarily chooses the pH values. Although it is important to interpolate and predict behaviour at pH values in between experimental points, there do not exist continuous processes in which the pH is evenly and gradually varied from one value to another unless the experimenter builds an apparatus to perform such an experiment.

However, continually varying sequences do occur regularly in chemistry. The most obvious processes involve the variation of one or more parameters with time: the parameter might be the proportion of an additive in a manufacturing process which needs to be monitored by methods such as quality control or the intensity of absorbance as a spectrum is scanned. Statistically these processes are formally called *time series*. However any *sequential analytical signal* can also be referred to as a time series. Another common example of a time series in chemistry is the measurement of a variable with distance, sometimes as a depth profile (e.g. the concentration of a chemical in a river) or as a geographic map (involving two dimensions). These processes can also be analysed using methods described below. A continually varying process in frequency (e.g. using a continually varying wavelength spectrometer) may also be considered as a time series.

In this chapter we will largely be concerned with the problems of sampling such series, although some consideration of the objective of the analysis will be necessary.

There are two principal reasons for sampling a time series. The first is *descriptive*. A model or description of the system is required. Under such circumstances if the sampling strategy has been poorly thought out it is frequently possible to return later and take more samples, or redesign the sampling scheme. Indeed sometimes sparse sampling can be used for exploratory analysis with more detailed sampling taking place later. There is one major difficulty with this approach in that material may be limited in availability. For example many geological cores are unique and expensive so once part of the core has been destroyed there is no chance of obtaining further information; mixing information from different cores can be disastrous. The second reason for sampling is to *monitor* a continuous process. This is especially important in quality control of manufacturing processes. If the quality of a product is poor then the manufacturing process must stop immediately. Thus rapid decisions need to be made: it is not possible to return to the system later to take more samples or reconsider crucial decisions. Sometimes major disasters can occur if the quality of a manufacturing process becomes out of control. Normally monitoring and *control* go hand-in-hand.

There is a large cultural gap between chemometricians who deal with experimental design and those who deal with time series analysis. This is probably because these are regarded as very different branches of statistics and have developed their own jargon, texts and terminology. Often many chemometricians involved in experimental design first encountered chemometrics *via* statistics for analytical chemistry whereas time series analysis, especially Fourier transform methods, is probably more relevant to spectroscopy.

It is a mistake to separate approaches for experimental design from approaches for sampling time series. Both types of situation occur commonly in chemistry and the true chemometrician must understand all types of problems and methods.

3.2 SAMPLING FOR MONITORING AND PROCESS CONTROL

Chemical processes such as manufacturing processes often change with time as factors vary: these factors might be fairly complex and could, for example, be humidity, operating temperature, pressure in reaction vessels, flow rate through pipes and so on. Experimental modelling of these parameters is often hard, and for real-time processes often impossible. It is easy to monitor the change in a parameter with time. This parameter might, for example, be the concentration of a chemical in tablets in the pharmaceutical industry: there is a large literature on this as there are often legal limits as to the composition of drugs, which manufacturers have to satisfy.

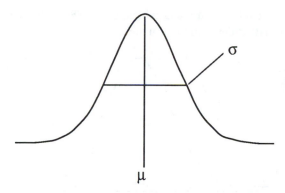

Fig. 3.1 - Gaussian distribution.

Often it is normal to measure the concentration of a chemical in a *batch* of products at regular time intervals. This concentration will have a mean $\bar{x}$ for each batch analysed. It is better to analyse batches rather than single tablets. With time this mean could change according to fluctuations in manufacturing process. Obviously if the process were perfectly behaved this value should remain fairly constant. However, the value of this mean will vary, and this variation can be described by a simple distribution, often a *normal* distribution. A normal distribution is also called a *Gaussian* distribution. The formula for this distribution is given by

$$y \quad = \quad \frac{e^{-(x-\mu)^2/2\sigma^2}}{\sigma\sqrt{2\pi}} \tag{3.1}$$

These distributions will be discussed in more detail in Chapter 5, but the main features of this curve are useful to note. Fig. 3.1 illustrates a typical Gaussian curve. The parameter μ represents the mean or centre of the curve. The standard deviation, which is proportional to the width at half height, is given by σ. The constant $1/\sigma\sqrt{(2\pi)}$ is a *normalization* constant so that the *area* under the curve adds up to 1. The main point to note is that there are three main parameters, due to position, width and area. These distributions are tabulated in standard statistical tables and can also be readily obtained from most standard software. It is usual to set a target area, for example α, which is either a number between 0 and 1 or 0 and 0.5 (the normal distribution is symmetric). For a two tailed normal distribution (α between 0 and 1) the tables will give z which is the distance away from the mean (μ) in units of σ, which divides the curve into three areas of $\alpha/2$, $1-\alpha$ and $\alpha/2$ as is illustrated in Fig. 3.2. Therefore the value of z for $\alpha=1$ will be 0. It is then possible to ask questions such as what is the region of the curve within which 50% of the readings are expected to occur? It is easy to verify, from standard statistical tables, that 50% of the readings should lie within 0.675 standard deviations from the mean.

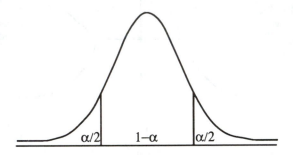

Fig. 3.2 - Areas of normal distribution that are usually tabulated.

How does this help analysing the quality of a product? The mean value of a parameter obtained from a batch is also likely to be normally distributed, so, if the process is behaving well, we expect the value of this parameter will have constant mean and standard deviation over time. For a normal distribution, 95% of the readings should be within 1.96 standard deviations of the mean, and 99.8% within 3.09 standard deviations. So only 1 in 1000 observations is expected outside approximately 3 standard deviations from the mean and 1 in 40 would be outside approximately 2 standard deviations from the mean if the process remained constant. These two limits ($\pm 2\sigma$ and $\pm 3\sigma$) are often called the *upper warning limit* and *upper action limit* respectively. If $\pm 2\sigma$ is exceeded it is probably worth investigating the behaviour of the system; if $\pm 3\sigma$ is exceeded then it is probably best to stop the process altogether. These limits are normally displayed graphically and this type of graph is often called a *Shewhart chart* (Fig. 3.3).

But as is usual in chemometrics, matters are not quite as simple as they appear at first. One major problem is how to estimate σ and μ, as this is clearly critical to the entire analysis. Often these parameters have to be estimated in advance on a very large sample. A second problem is how large a sample should be monitored in each batch: a typical area where quality control is important is in estimating the chemical constituents of a batch of tablets during a production process. If 1000 tablets are manufactured per hour, what size sample is used to estimate the batch mean? It is not efficient or cost effective to use a sample of 100, but should we use 20 or 5 tablets? The size of the batch monitored during the process will almost certainly be less than the size of the sample used to estimate σ and μ. However, it is normal to use only a single parameter, the mean of each batch, for monitoring the process. Because several measurements are averaged per batch, this expected batch standard deviation will be less than the standard deviation of individual samples and is given by

$$\sigma' \quad = \quad \sigma/\sqrt{N} \qquad (3.2)$$

where N is the number of samples in a batch and σ' is the standard deviation of the batch mean, so the axis of Fig. 3.3 is in units of σ'.

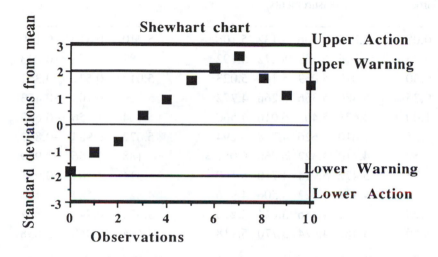

Fig. 3.3 - Shewhart chart with action and warning limits indicated.

Sometimes the process might get out of control because the variability in the sampling increases: this is equivalent to the Gaussian distribution changing in width rather than position. Under such circumstances it is worth looking at the distribution of measurements within a batch, rather than the mean measurement per batch. A parameter w is defined as the range of measurements, i.e. the difference between the largest and smallest measurement in each batch. This similarly can be displayed graphically in a quality control chart (QC) chart. There are various standard methods for computing the control limits, which are usually taken care of by standard QC software packages.

Another way of monitoring quality is by using a cumulative summation of the difference between a parameter and the theoretical mean, i.e.

$$c \quad = \quad \sum_{i=1}^{i=I} (\bar{x}_i - \mu) \qquad (3.3)$$

where μ is the overall expected mean, and $\bar{x}_i$ is the mean for a given batch i and I the number of batches sampled. If the process is under control, this value should remain close to 0. It is often obvious graphically when this process has got out of control. The parameter c is often called the *cusum* (cumulative summation).

Table 3.1 - Quality control example

Time	Measurements				$\bar{x}$	w	c
0.00	5.096	5.506	5.132	5.106	5.210	0.410	−0.110
0.25	4.716	6.369	5.272	5.192	5.387	1.653	−0.043
0.50	4.819	5.039	5.165	5.023	5.012	0.571	−0.351
0.75	5.480	5.356	5.266	4.772	5.219	0.708	−0.453
1.00	4.678	5.402	6.016	4.560	5.164	1.456	−0.609
1.25	5.610	4.626	4.758	5.294	5.072	0.984	−0.857
1.50	4.310	4.322	6.066	6.052	5.188	1.756	−0.989
1.75	5.472	5.914	4.710	5.160	5.314	1.204	−0.995
2.00	4.888	3.922	5.506	4.986	4.825	1.584	−1.490
2.25	3.672	4.386	3.644	4.210	3.978	0.742	−2.832
2.50	4.462	4.624	3.770	5.118	4.494	1.348	−3.658

Note : The theoretical mean for the process is $\mu = 5.32$. Each batch consists of four measurements.

The calculation of $\bar{x}$, w and c is illustrated in Table 3.1 for a dataset, together with a graph of the cumulative summation in Fig. 3.4 for this process. It is obvious from the graph of c that the process is going out of control. This is not so obvious, at first glance, from the actual measurements. For example the mean measurement at time 0 (5.210) is actually less than the mean measurement at time 1.75 (5.314). However, both measurements are less than the expected mean ($\mu=5.32$).[1] The value of w obviously is likely to increase the larger the batch. Clearly the process would be monitored better by choosing a larger batch but there often is not the time available to analyse large batches. Naturally the analysis would differ if the overall population mean (5.32) differed. The value of μ is computed prior to monitoring the process and, is, of course, critical to the entire analysis.

[1]The numbers were in fact simulated by the equation $5.32t \, e^{-0.05(t^2-t)} + 2 \, (v-1)$ where t is the time, and v is a random number generator giving a uniform distribution between 0 and 1 (so there is equal chance of obtaining any value $0 \le v \le 1$).

The expected mean and standard deviation of a process is not always known in advance. Some processes cannot be studied in such detail. In these cases, a

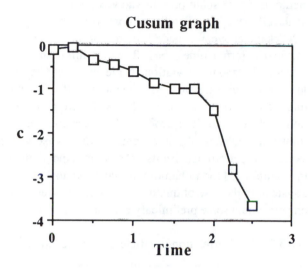

Fig. 3.4 - Cusum graph of data in Table 3.1.

running mean or standard deviation can be computed, or a weighted parameter dependent on the most recent runs. In fact there are a large number of options in QC charts and a vast statistical literature, so the chemometrician should think very carefully as to what main parameters are required from the system and how best these can be monitored. Decisions such as the rate of sampling, the number of samples in a batch, how and whether initial control limits should be estimated, and the type of QC chart to be used need to be taken in advance of experimentation. Incorrect decisions can lead to disastrous results. Sometimes a conservative estimate of control limits is required, particularly if a major disaster would result from a poor product. In fact in regulated industries such as the pharmaceutical industry there often is a legal requirement that concentration of compounds in a drug is within certain limits: one poor batch could lead to prosecution. In other industries variations in the quality of a product might influence taste or appearance of a product and not really lead to a disaster, so much time could be wasted investigating "false alarms"; therefore rather less sensitive tests are required.

In all the schemes outlined above sampling was performed at *regular* time intervals. In fact this approach is rather inefficient. If a process is behaving well (i.e. close to the theoretical mean), there is no need to sample rapidly. If the process appears to be getting out of control, more rapid sampling is required. In this way costly analyses (often requiring reagents or manpower or instruments) can be minimized.

Much of chemometrics is about putting knowledge of a system to work, in order to increase the efficiency or reliability or cost effectiveness of a process. In the analysis of QC charts above, we made little use of the fact that the readings are related in time and so form a time series. The only information we used was the predicted mean and standard deviation of the process. Clearly a "better" approach would be to examine the process in even more detail. Of course, like statistical optimization described in Chapter 2, how far we go depends on how much time is available to study the process. The best method of sampling depends on complete knowledge of the system, in which case we can predict when it will exceed predetermined limits in advance and there is no point carrying out any sampling. Good chemometricians use to advantage *partial knowledge* of a system. In the case of quality control of a process with time, this might depend on performing some preliminary experiments.

Table 3.2 - Autocorrelation coefficient : simple example

Time	x_0	x_1	x_2
0	4.407	4.529	4.486
1	4.529	4.486	4.578
2	4.486	4.578	4.517
3	4.578	4.517	4.445
4	4.517	4.445	4.593
5	4.445	4.593	4.570
6	4.593	4.570	4.594
7	4.570	4.594	4.563
8	4.594	4.563	4.702
9	4.563	4.702	.
10	4.702	.	.

An important parameter when studying time series is called the *autocorrelation function*. The correlation coefficient was defined in Eq. (2.1): for a variable correlated with itself, this coefficient will always equal exactly 1. Consider, though, a time series consisting of N points. In addition to the correlation coefficient of the time series with itself, which we will call ρ_0, it is possible to shift the time series by a given amount and then recalculate the correlation coefficient of the time series with itself. So for Table 3.2, ρ_0 is the correlation coefficient of x_0 with x_0, ρ_1 is the correlation coefficient of x_0 with x_1, ρ_2 is the correlation coefficient of x_0 with x_2 and so on. The graph of ρ_i versus i can be displayed, which for the data in Table 3.2 is given in Fig. 3.5: the values of ρ_i are tabulated in Table 3.3. This graph is commonly called an *autocorrelogram*. In Section 7.8 we will discuss the importance and properties of

Table 3.3 - Autocorrelation coefficients for data in Table 3.2

i	ρ_i
0	1.00
1	0.11
2	0.29
3	0.18
4	0.34
5	0.20
6	−0.05

the autocorrelogram in greater detail, but it is worth noting at this stage that these graphs contain a great deal of diagnostic information about time series. The formula for the autocorrelation coefficient is given by

$$\rho_i = \frac{[\sum_{t=1}^{N-i} (x_t - \bar{x}_{N-i,1})(x_{t+i} - \bar{x}_{N,i+1})] / (N-i)}{[\sum_{t=1}^{N} (x_t - \bar{x}_{N,1})^2] / N}$$ (3.4)

where t is an integer representing the tth sequential readings, there are N points in the original dataset, and ρ_i is the correlation coefficient as the time series is shifted by i positions against itself; $\bar{x}_{M,m}$ is the mean of readings m to M. The value of i is often called the *lag*, and is in units of the time interval. An alternative formula for Eq. (3.4) is given by

$$\rho_i = \frac{((N-i) \sum_{t=1}^{t=N-i} x_t x_{t+i} - \sum_{t=1}^{t=N-i} x_t \sum_{t=1}^{t=N-i} x_{t+i})/(N-i)^2}{(N \sum_{t=1}^{t=N} x_t^2 - (\sum_{t=1}^{t=N} x_t^2)^2)/N^2}$$ (3.5)

This latter formula may look complicated, but is far easier to compute than Eq. (3.4) and provides an identical answer. Note that the autocorrelation coefficient takes a value between −1 and +1 in the same way as a correlation coefficient.[1]

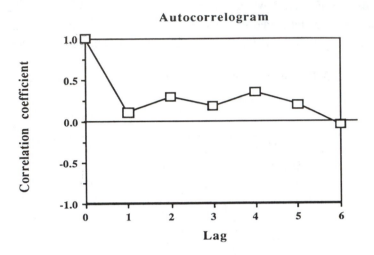

Fig. 3.5 -Autocorrelogram from Table 3.3.

One of the most important properties of an autocorrelogram is that it reduces noise in the data. A simple type of noise is called *white noise*: the value of noise at each successive reading is unrelated to the noise at the last reading. It is simple to simulate such noise using a random number generator. If the noise free value of the response at time t is given by $\hat{x}_t$, then

$$x_t = \hat{x}_t + g \qquad (3.6)$$

where g is a noise function. Since this type of noise is unrelated in time, in other words, the value of noise at time t is independent of the value of noise at time $t+1$, if we simulate the noise using a random number generator, we just add on the random number to the theoretical value of the time series at each sampling point. We will see that this approach is quite a useful description of spectroscopic noise but is not a very accurate description of noise in naturally occurring time series or in manufacturing and other processes. More usually, the noise is *coloured noise*: that means that the value of noise at each time interval is related to the noise at the last time interval. Statisticians have studied these noise

---

[1] Note that some authors use $(N-i)(N-i-1)^2$ instead of $(N-i)^2$ and $N(N-i)$ instead of N^2 in the definition of a correlation coefficient (Eq. (3.5)). For long series and small lags this makes very little difference. The reason is that some authors prefer to use sample rather than population standard deviations: in such cases the residual sum of squares of a set of numbers is divided by $(M-1)$ where M is the number of readings in the dataset, rather than M.

distributions in some detail. The classical work of Box and Jenkins is essential reading for the specialist in this area, but, briefly, noise is modelled by a series called an *autoregressive moving average* or ARMA process. In such a process successive noise values are related to previous values of noise. This is quite a reasonable model for noise in many systems. A simple process might be described by the following equation

$$v_t \quad = \quad v_{t-1} + g \qquad\qquad (3.7)$$

in which v_t is the noise at time t and g is, again, a random function. So if the noise at time $t-1$ is +1 then, if the noise function is symmetric, a noise level of +0.5 or +1.5 is more likely at time t than a noise level of 0 or −0.5, for example. Much more complex noise functions can be proposed, although often the chemometrician does not have time to extensively study the noise in the system. However, if there is coloured noise in the system, this means that the noise itself will form a sort of time series.

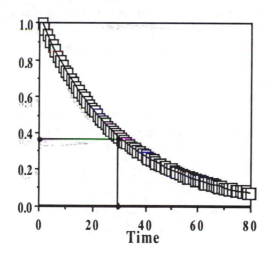

Fig. 3.6 - Estimation of the time constant of an exponential decay.

If there were no noise in a system, and the process remained absolutely steady over time, the autocorrelogram would be a straight line of constant amplitude 1. Because of noise and system variability, this does not occur, and if there is an element of coloured noise, a non-cyclic process normally exhibits a decaying autocorrelogram that can be approximated by a decaying exponential, so that value

$$\rho_t \quad = \quad e^{-t/T_x} \qquad\qquad (3.8)$$

The parameter T_x is a *time constant*. There are various ways in which this constant can be estimated, but one method is to pick the value of t at which the autocorrelogram has decayed to $1/e$ or 0.368 of its original value: it is easy to verify that this value is identical to T_x for a pure exponential of the form of Eq. (3.8). This is illustrated graphically in Fig. 3.6. Clearly the greater this value the more slowly decaying the exponential.

So a time constant T_x can be determined for a given process in addition to σ and μ: we now have further information at our fingertips. This now can be used to give some idea as to how fast the process is likely to get out of control. The smaller the time constant, the faster the process could get out of control. Various other parameters can be built into the system. The standard deviation of the errors can be broken down into the standard deviation of the analytical error (analogous to the analytical error discussed in Chapter 2) (σ_a) and the standard deviation of the underlying process (σ_t) (analogous to the lack-of-fit error). Another parameter that can be used is the *reliability factor*. The higher this factor the lower the chance of error in the predicted response at time $t+1$. To reduce the chance of error to 0, it is obviously necessary to sample infinitely fast, which is clearly impossible. If the threshold value (or warning / action limit) is given by M in units of σ_t[1] then the reliability factor, R, must be greater than M. Typically a reliability factor of $R = M + 1.5$ will provide a 99% chance of being correct that a predicted response is within the limit of M.[2]

So it is necessary to sample sufficiently frequently to ensure we have high confidence that each successive sample is within the set control limits. An equation for the time between one sample and the next is given by

$$\tau = -T_x \ln\frac{[M\,x_t + R\sqrt{x_t^2 - q\,(M^2 - R^2)}]}{(x_t^2 - q\,R^2)} \qquad (3.9)$$

where x_t is the value of the response sampled at time t and the next sample is to be taken at time $t+\tau$, and $q = (\sigma_t^2 + \sigma_a^2)/\sigma_t^2$, which equals 1 if analytical error is negligible.

Of course more elaborate and even more "efficient" formulae can be proposed, but where do we stop? The problem of how frequently to sample time series is typical of many chemometric situations. The more that is known about the system (e.g. the time constant of the autocorrelogram of the process) the fewer samples are required, but the more prior knowledge is required. It is not very useful to pull a formula out of a textbook (even out of this book!) without some understanding of the physical background behind the derivation, and if the chemometrician does not fully understand the assumptions behind a complex (but supposedly labour saving) method, he is probably best placed using a

[1] Typically 2 or 3 standard deviations as discussed above, but any value, can, of course, be employed.
[2] See paper by Muskens in the bibliography.

simpler approach. The more that is known about the nature of the noise and signal, though, the more sophisticated the method that can be used. This has practical considerations. For example, it might be useful to issue warnings when a process is beginning to go of control. If these thresholds are too liberal, warnings might be issued when, in practice, the process is behaving reasonably, but, because there is insufficient information available as to the optimum sampling rate or the nature of noise in the system such warnings are necessary in order to prevent accidents. In order to set up a more realistic threshold, the system needs to be studied in far greater depth, which can take time, expertise and manpower. The laboratory manager must carefully choose his method.

3.3 SAMPLING FOR DESCRIPTION

A different reason for sampling is to describe a system. Such a system might be a rock or a riverbed. The concentration and distribution of chemicals or related parameters (e.g. colour, texture) in the system may be of interest. How should the material be sampled and how much material should be taken per sample?

This is a problem of *experimental design*. In Section 3.4 we will discuss cyclical processes, but in this section we will be concerned with processes that can be described by polynomial functions, in a similar way to the processes described in Chapter 2. One major difference, though, is that the processes discussed here are best described by *local* polynomials, that is, that the nature of the polynomial changes throughout the region of response. These polynomials are often also described as *moving averages*. The simplest form of linear moving average is given by

$$\hat{x}_0 = \sum_{n=1}^{N} \alpha_n x_n \qquad (3.10)$$

where α_n are coefficients, and x_n are readings at N sampling points. Note that the estimated value of x does not need to be at a sampling point, but can be an intermediate point. Indeed, as explained in Chapter 2, in many cases of interest, estimated points will not correspond to experimental sampling points, since a major objective of experimentation is to discover the value of a response at points that have not been sampled, and to build up a description of a system using the minimum number of samples necessary. Moving averages will be discussed in greater detail in Chapter 5. The objective of a good descriptive sampling strategy must be to obtain reliable values of $\hat{x}$. How reliable these values are and how many sampling points are required to build up an adequate picture can be determined using chemometric approaches.

In describing a continuous system it is normal to start with some sort of model of the data. An approximate theoretical model of variability is usually employed. This involves computing a *variogram*. The *semivariance* is defined by

$$\gamma_m = \frac{1}{2N}\sum_{n=1}^{N-m}(x_n-x_{n+m})^2 \tag{3.11}$$

Table 3.4 - Data for variogram example

n	x	$m=1$ $(x_n-x_{n+1})^2$	$m=12$ $(x_n-x_{n+12})^2$
1	−0.629	0.192	2.934
2	−0.191	0.130	1.464
3	0.170	0.083	0.571
4	0.458	0.050	0.118
5	0.681	0.028	0.001
6	0.848	0.015	0.154
7	0.969	0.007	0.549
8	1.053	0.003	1.188
9	1.107	0.001	2.097
10	1.135	0.000	.
11	1.141	0.000	.
12	1.124	0.002	.
13	1.084	0.004	.
14	1.019	0.009	.
15	0.926	0.015	.
16	0.802	0.024	.
17	0.646	0.036	.
18	0.455	0.052	.
19	0.228	0.070	.
20	−0.037	0.092	.
21	−0.341	.	.

$$\sum_{i=1}^{21-j}(x_i-x_{i+1})^2/42 \qquad\qquad 0.0194 \qquad\qquad 0.2161$$

Note : Calculations for *m*=1 and *m*=12 shown in full.

where x_n is the value of the response at position n in a sequence of N readings. The semivariance for $m=0$ is equal to 0. Eq. (3.11) assumes a regularly sampled sequence. The value of γ_m is then an average variability at m points away. This parameter might seem somewhat complex, so we will illustrate this with a simple numerical example.

Table 3.4 lists 21 readings, equally spaced in time, of x_n.[1] The values of $(x_n-x_m)^2$ are computed for $m=1$ and $m=12$. Obviously they may be calculated for any other value of m. The resultant values of γ_m using the formula in Eq. (3.11) are also computed in Table 3.5.

Table 3.5 - Values of the semivariance from data in Table 3.4

m	γ_m	γ_m/σ^2
0	0.000	0.000
1	0.019	0.069
2	0.063	0.226
3	0.115	0.411
4	0.165	0.590
5	0.207	0.740
6	0.239	0.854
7	0.259	0.928
8	0.269	0.962
9	0.269	0.961
10	0.259	0.926
11	0.241	0.862
12	0.216	0.773
13	0.185	0.663
14	0.151	0.540

Another important property of the semivariogram is demonstrated in Table 3.5. The overall variance of the readings as a whole is 0.280. It can be shown that for large values of m, the value of γ_m converges on the variance of the original readings. The value of γ_m/σ^2 is also computed in Table 3.5, using the definition in Eq. (3.11). It is indeed seen to converge. As $m>8$ the value of this parameter falls slightly. This is a result of the small sample size. If $m=20$ the

[1] The readings were, in fact, computed using the formula $x_i = 0.25(i/4) - 0.32(i/4-3.6)^2 + e^{(i/4)/2-(i/4)^2/4}$ as a simulation dataset, but the reader can use any sequential dataset.

semivariance is totally dependent on the difference between the extremes. However, the change in this parameter which is exactly analogous to a correlation coefficient can be plotted as a graph (Fig. 3.7). It is possible to use the square of this parameter if preferred. The graph in Fig. 3.7 is called a variogram.[1]

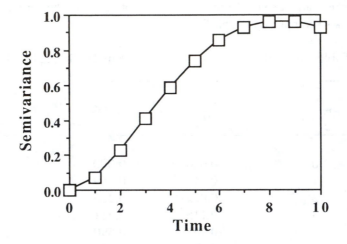

Fig. 3.7 - Variogram from Table 3.5.

The shape of the variogram, which theoretically varies between 0 at $m=0$ and 1 at $m=\infty$, helps in the design and analysis of a good spatial sampling strategy. Often a theoretical model of a form such as given in Eq. (3.12) is fitted to the data

$$\gamma_m = \sigma^2 (1 - e^{-\lambda m}) \qquad (3.12)$$

but there are several other possible approaches in the literature. Although there are several empirical ways of establishing such models, it is normally relatively easy to take several equally spaced samples, calculate the semivariogram and fit a model such as the one of Eq. (3.12).

Sometimes, we record samples in more than one dimension. The use of semivariances for the analysis and design of two or more dimensional sampling experiments is generally called *Kriging* and was originally developed by geologists, but has implications for all sequential chemometric experiments. With a knowledge of the semivariogram it is possible to design a sampling strategy in which N sampling points are used to estimate the value of the

---

[1]Note that some authors divide by $(N-m)$ in Eq. (3.11) rather than N.

parameter of interest at another sampling point. A typical design involves eight sampling points on the corners and edges of a square (Fig. 3.8). This involves performing eight experiments to estimate the value of the parameter in the centre of the square.

Kriging can be used to fit a moving average, so that

$$\hat{x}_0 = \sum_{n=1}^{N} \alpha_n x_n \qquad (3.13)$$

where $\hat{x}_0$ is the estimated value of x in the centre of the experiments. The values of α_n can be calculated using the equations

$$\sum_{n=1}^{N} \alpha_n \Gamma_{nm} = \Gamma_{m0} - \lambda \qquad (3.14)$$

and

$$\sum_{n=1}^{N} \alpha_n = 1 \qquad (3.15)$$

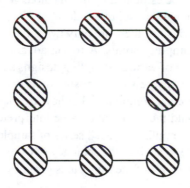

Fig. 3.8 - Typical sampling strategy for Kriging. The eight points are sampled and used to provide an estimate of the parameter in the middle of the square.

There will be N equations of the form of Eq. (3.14): each value x_n is the observed value of the parameter (e.g. the concentration of a chemical) at each sampling point, used to estimate the value of the parameter at the central sampling point. These equations could also be expressed in matrix form. For the design of Fig. 3.8, $N=8$ and there will only be two different values of α, since the eight sampling points form two geometric groups, namely those at the corners and on the edges of the square. The condition of Eq. (3.15) is used to normalize the coefficients. The parameter Γ_{nm} is the value of the theoretical semivariance between points n and m. It is related to γ_n in since it is a function only of the distance between the two sampling points. Γ_{m0} is the value of the theoretical semivariance between the point of interest and sampling point m. The value of λ is an unknown but can be calculated from Eqs (3.14) and (3.15) providing the semivariance is already known. This analysis assumes that the material is *isotropic*, that is the semivariogram is roughly the same in any direction (in two or more dimensions). If this is not so the analysis becomes considerably more complex and is out of the scope of this text. In fact probably a more complex sampling strategy is required.

Perhaps the reader may be asking why he or she should not be directly sampling the estimated point and what the advantages are of using such an elaborate procedure. There are various reasons. The most important is that there is always noise in a system. By using the values of the observed parameters in nearby sampling points it is possible to get a more accurate estimate of the value of the parameter x_0. Effectively we use a moving average smoothing function: estimates of parameters are often more accurate than individual experimental observations. The second important aspect of the analysis is that it really is not too crucial if there are slight errors in the model of the semivariance: the analysis is better than using linear interpolation. Third, the value of the parameter at any point on the surface can be estimated by this method. More symmetric strategies result in easier analysis, of course. In the design in Fig. 3.8, only two values of α need to be determined, since there are only two different distances between the centre point and the sampling points. A fourth advantage is that this approach allows detailed mapping of various sampling designs such as grid designs. This could be important in many geochemical situations, for example. The resultant response may be a principal component, but how frequently should the samples be taken and how should the resultant data be interpreted? Maps of the region can be computed from a regularly spaced series of samples.

Finally it is possible to analyse the accuracy of estimation of a response. The variance or mean square error of the estimate is given by

$$s^2 = \lambda + \sum_{n=1}^{N} \alpha_n \Gamma_{n0} \tag{3.16}$$

The greater the distance between the sampling points and the estimated point relative to the semivariance the larger this error, as Γ_{m0} is greater the farther the sampling point m is from the estimated point, since the value of the semivariance increases with distance. Thus the error in estimated values depends in part on sampling strategy. Therefore, with some knowledge of the shape of the variogram it is possible to design a sampling strategy taking into account the tolerable size of errors in the system.

Although Kriging is not formally considered a form of experimental design, in practice it can be used to build models and design sampling strategy for one, two and three dimensional processes. If time is taken into account, four dimensional Kriging is possible. There are a number of other approaches to the design of sequential sampling strategies to describe a system but these have been applied less to chemical systems so will not be described in detail here.

3.4 SAMPLING SEQUENTIAL TIME SERIES

Some time series contain cyclical information. Typical examples are Fourier series and naturally occurring time series such as occur in geochemistry and environmental chemistry. In this case the elements of interest are normally a sum of sine waves. The time series can be described by

$$f(t) = \sum_{j=1}^{J} \cos(t\zeta_j + \phi_j) + \nu_t \tag{3.17}$$

for samples at time t, and J components (each described by a sine wave of frequency ζ_j and phase angle ϕ_j) and a noise function ν_t. The direct handling of Fourier series is described in greater detail in Chapter 5. The noise function may contain systematic components. The sine waves may reflect the change in concentration of a compound with time, or a spectroscopic component.[1]

It is important to consider how frequently to sample such series. There are two situations. The first is when it is always possible to sample series regularly and, effectively, obtain infinitely accurate samples. This case is usually the situation in spectroscopy. For example, in NMR magnetization can be sampled regularly in time, and the main consideration is how much disc space is available and what acquisition time is required. A second case is when there are serious sampling difficulties. This may occur when sampling naturally occurring time series such as are found in geochemical or environmental studies. Often there is not sufficient material for indefinitely frequent sampling. Sometimes the size of sample has to be large relative to sampling intervals in order to obtain sufficient sample for sensitive analytical instruments.

The case of regular sampling is easiest to analyse. The maximum range of uniquely distinguishable frequencies is given by the *Nyquist frequency*. This is best illustrated graphically (Fig. 3.9). If a series is sampled once per δ s, then a component oscillating at a rate of $1/\delta$ cycles s^{-1} will appear to be oscillating at 0 cycles s^{-1} as it will appear identical to a zero frequency component. A component oscillating at $(1/2\delta + \Delta)$ cycles s^{-1} will appear identical to one oscillating at $(1/2\delta - \Delta)$ cycles s^{-1}. Thus $1/2\delta$ cycles s^{-1} is the *maximum distinguishable frequency* and is called the *Nyquist frequency*. In a Fourier transform components actually oscillating above the Nyquist frequency are reflected in *foldover*, so components at high frequencies will appear in identical

---

[1]In this text we will use units of cycles s^{-1} to describe frequencies, rather than radians s^{-1}. This unit of measurement is discussed in greater detail in Section 5.1.

Sampling rates and Nyquist frequency

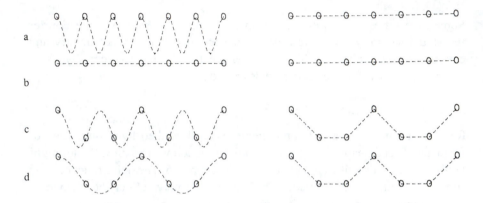

Fig. 3.9 - Demonstration of Nyquist frequency. The true and observed sinewaves
are illustrated.

Demonstration of foldover

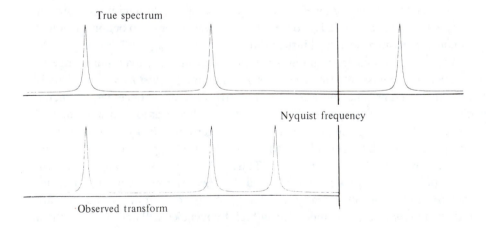

Fig. 3.10 - Foldover of high frequency components.

positions as lower frequency components (Fig. 3.10). The way to increase the range of observable frequencies is to increase sampling frequency. However, this solution is not quite as simple as it seems. Normally the overall datalength of a time series is limited; this is because computers and disc files have limited memories or because each sample involves work to acquire and manpower is limited. So, if the sampling rate is doubled, but the time series is sampled for the same overall time, twice as much data needs to be acquired. If this is impracticable and the number of datapoints is kept the same (so that, for example, 1000 datapoints are acquired over 0.5 s rather than 1 s, therefore doubling sampling rate without increasing the number of datapoints sampled), the *digital resolution* decreases. That is because, in order to double the range of frequencies, but keep the number of datapoints constant, the time series is only sampled half as rapidly.

Substantially greater difficulties occur when sampling naturally occurring time series. These may occur, for example, in geochemistry or environmental chemistry. Cyclical changes such as those associated with diurnal and tidal motions (environmental chemistry) or long terms changes of the orientation of the earth's orbit around the sun (geochemistry) often have an influence on the change of concentration of chemicals in material such as water samples or geochemical sediments. Unfortunately there are often serious difficulties in regular sampling of material. Consider the problem of a geological core. It might be desired, first, to sample regularly in time. However, time and depth are not necessarily linearly correlated. This is because material is deposited unevenly in time. When sampling a core the depth / time calibration curve might not be known in advance. Therefore sampling evenly in depth does not guarantee even samples in time. The errors may be quite serious and, worse, systematic. A second problem is if there simply is not the material or manpower to sample completely evenly. For example consider the problem of manual sampling of river water with time. It might not be possible to sample at exact intervals because of limited manpower or even lunchbreaks.

These problems may be overcome by *interpolation* to a regularly sampled series. There are a number of interpolation methods. The simplest is linear interpolation. To calculate the estimated value of x and time t ($\hat{x}_t$), then

$$\hat{x}_t = \frac{(t_{+1}-t)x_{t-1}+(t-t_{-1})x_{t+1}}{(t_{+1}-t_{-1})} \qquad (3.18)$$

where t_{-1} and t_{+1} are sampling times immediately before and after time t (if a point is sampled exactly at time t then no correction is necessary). This is illustrated in Fig 3.11. This method of interpolation can be extended to a weighted interpolation, and can take into account any number of sampling points to give

$$\hat{x}_t \quad = \quad \sum_{i=-j}^{i=j} \lambda_i x_{t_i} \tag{3.19}$$

where λ_i is a weighting factor. This approach allows for smoothing and reduction of noise as well as interpolation. These weighting functions are described in more detail in Chapter 5, but it is useful to recognize that smoothing and interpolation can be combined in one step. Obviously interpolation functions do not need to be linear. Indeed the more that is known about the system the more elaborate the interpolation method. Of course, if everything is known about a system, there is no point processing the data further, so the chemometrician normally needs to make informed guesses about the nature of signals and noise before using elaborate methods. If very little is known about a given system, a simple linear approach, such as that of Eq. (3.18), is probably best. The more sophisticated the method of interpolation the more assumptions about the system are, in practice, built into the model. If these assumptions are incorrect, serious errors can occur in the resultant reconstruction. On the other hand if there is certain knowledge about the nature of a system this should be incorporated if possible.

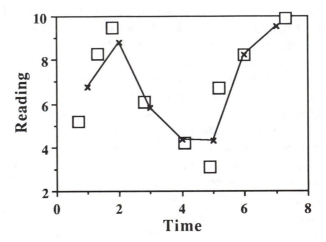

Fig. 3.11 - Data linearly interpreted according to Eq. (3.18) to uniform intervals.

Another problem is size of sample (or lot). Instrumental methods such as GC might be used to determine the concentration of chemical in each sample. The smaller the sample, the lower the signal to noise ratio, and, therefore, the less reliable the resultant quantitation. Thus sample sizes (e.g. length of a geological core) may need to be close to sampling frequency. Fig. 3.12 illustrates a sampling strategy where sample sizes need to be about half the sampling rate to provide adequate sample size. This introduces further errors. Normally if material is sampled between, for example, 30 and 39 cm in depth, then the average value of a parameter measured over that distance is set at 35 cm. However, if the true value of the parameter (e.g. the concentration of a chemical) varies non-linearly over such a period, the average over this period will not correspond to the true value. The magnitude of such an error is best illustrated by a simple numerical example. Consider trying to estimate the value of $\cos\alpha t$ when sampling over the period between $t-\delta t$ and $t+\delta t$. This estimate is given by (all angles are in cycles s^{-1})

$$\hat{x} = \frac{\int_{t-\delta t}^{\tau=t+\delta t} \cos\alpha\tau d\tau}{2\delta t} \qquad (3.20)$$

where $x = \cos\alpha t$. It is simple to show that this becomes

$$\hat{x} = \frac{\sin\alpha\delta t \cos\alpha t}{2\pi\alpha\delta t} \qquad (3.21)$$

As $\delta t \rightarrow 0$, this expression $\rightarrow \cos\alpha t$, as expected.

Table 3.6 - Effect of sample size of estimated values of a sparsely sampled cyclical process

Time	δt			
	0	1	3	5
5	0.203	0.200	0.181	0.146
15	−0.577	−0.570	−0.514	−0.413
25	0.854	0.843	0.762	0.612
35	−0.991	−0.978	−0.883	−0.710
45	0.963	0.951	0.859	0.690
55	−0.776	−0.766	−0.692	−0.556

Note : A value of time = 25 and δt=3, would imply that the sample is sampled between time = 22 and 28 units, so the value recorded for time = 25 is, in fact, the average over that period. A sinecurve of the form $y = \cos(t/23)$ has been simulated. The values for δt=0 are the exact values at any given time. Note that frequency is in units of cycles s^{-1}.

Table 3.6 illustrates the calculation of the errors for a curve of the form cos $(t/23)$ for $\delta t = 0, 1, 3, 5$ and sampling every t=10 time units.[1] It is easy to

[1]Note that this implies that the value of the sinecurve for t=23 is 1.

compute similar data for any combination of values of α and δt, and sampling interval. However, from Table 3.6 it should be clear that the observed values are closer to 0 than the true values of x. The greater δt, the closer these are to zero. This is intuitively expected as the mean value of a cosine wave is 0. The distortion is greater the higher the frequency of the component, i.e. peaks closer to the Nyquist frequency are distorted more than low frequency components. This problem of sample sizes can result in a very serious and systematic error in estimates of intensities of components and careful thought, therefore, must be given to sampling errors if sample size in time (or depth etc.) is of the same order of magnitude as sampling frequency.

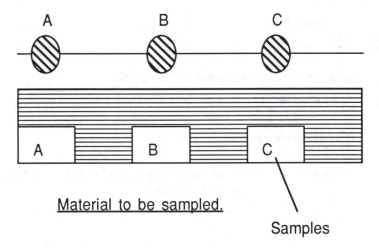

Fig. 3.12 - Large sample size or lot. The samples A, B and C are large relative to the amount of material available, but are used to estimate discrete values of a parameter.

Possibly the only realistic way of estimating such errors is through simulation of the particular data with some knowledge of the actual sampling errors. It is impossible to provide general rules without looking at the specific problem under consideration. In addition to some guess as to the type of cyclicity expected in the data and sampling problems, knowledge of the noise distribution can also help. In many cases data close to the Nyquist frequency are substantially distorted and attempts to deduce quantitative models and information in such situations should be treated with some caution. Sometimes thought should be given to the overall experimental approach. Often a time series can be sampled more frequently and more regularly if more sensitive analytical techniques are employed. For example, if one technique has a signal to noise ratio 10 times that of another technique then equivalent information can be obtained from samples one tenth in size. Another possibility is to use replicates analysis (if possible) to cut down the sample size. Careful thought and planning must go into working out an optimal sampling strategy for cyclical time series.

3.5 SAMPLING IN THE CONTEXT OF CHEMOMETRICS

There is now a considerable body of literature on experimental design in chemistry and a growing awareness of the need to design experiments prior to multivariate analysis. Much less attention has been placed on sampling strategies for sequential series in chemistry, yet these occur in many situations whether continuous monitoring of manufacturing processes or the study of natural cyclical processes.

In this chapter we have concentrated almost exclusively on univariate time series. However, there is no reason why the observed parameter needs to be a single measurement: it may be a suite of measurements, e.g. several HPLC peaks or AA (atomic absorption) measurements. With the advent of modern computerized instruments these types of multivariate sequential measurements become more possible. Chemometric methods for the design of sampling strategy will aid the experimenter considerably in the future and are an important, but often neglected, area of chemometrics.

REFERENCES

General
G.Kateman and F.W.Pijpers, *Quality Control in Analytical Chemistry*, Wiley, New York, 1981
G.Kateman, *Chemometrics Int. Lab. Systems*, 4, 187 (1988)
B.Kratochvic, D.Wallace and J.K.Taylor, *Anal. Chem.*, 56. 113R (1984)
J.C.Davis, *Statistics and Data Analysis in Geology, Second Edition*, Wiley, New York, 1986

Section 3.2
J.C.Miller and J.N.Miller, *Statistics for Analytical Chemistry, Second Edition*, Ellis Horwood, Chichester, 1988
O.L.Davies and P.L.Goldsmith, *Statistical Methods in Research and Production*, Oliver and Boyd, Edinburgh, 1972
P.J.W.M.Muskens, *Anal. Chim. Acta*, 102, 445 (1978)
G.E.P.Box and G.M.Jenkins, *Time Series Analysis, Forecasting and Control*, Holden-Day, San Francisco, California, 1970

Section 3.3
J.C.Davis in B.R.Kowalski (Editor), *Chemometrics : Mathematics and Statistics in Chemistry*, Reidel, Dordrecht, 1984, p. 419

Section 3.4
R.G.Brereton, *Chemometrics Int. Lab. Systems*, 1, 17 (1986)

4

Choosing and optimizing analytical conditions

4.1 ANALYTICAL METHODS

Chemical information is frequently estimated indirectly. Measurements such as electrical current, magnetic field strengths, time are converted to chemically interesting information such as concentrations, spectroscopic intensities, pH and so on. Yet there is another layer to the process of answering chemical questions. Frequently the questions asked are qualitative, such as is a river polluted? or does a cheese come from a certain region? Concentrations, chromatographic intensities, spectroscopic frequencies and so on are used to answer these qualitative questions. So the chemist may record a series of voltages and from these deduce whether a river is polluted or not. Chemometrics helps the chemist make best use of this information. In Chapters 2 and 3 we discussed the problems of design of sampling strategies. Once a sampling strategy has been decided upon, it is necessary to choose and optimize the measurement system. Although we might decide upon an excellent sampling strategy for recording the concentration of a chemical with time we then have to choose how (physically) we estimate this concentration. In fact the measurement process cannot be separated from sampling strategy. The better the measurement process the less noise there is in the samples and the easier it is to fit a model or the smaller the sample size required. An important feature of chemometrics, which is heavily dependent on the revolution in laboratory instrumentation, is that it is possible to choose which aspect of the experimental and analytical process can be improved most profitably: the chemometrician needs considerable broadness of mind in looking at the entire problem as a whole.

A great deal of chemometrics involves the optimal management of resources. The chemometrician might like to ask whether a river is polluted or not. How does he sample the river? What techniques are needed to analyse the river water? How is the quality of the instrumental data improved? How are the resultant data analysed? It is impossible to provide exact answers to these problems. A great deal depends on what resources are available. What manpower is available? What training have people had? Are there statisticians or experimental technicians on the payroll? How much effort is worth putting into the problem? What instrumentation is available? How much funds are available to hire extra people or purchase extra instruments? How much time is it worth investing in trying to answer these questions?

In this chapter we will look at ways of choosing and improving measurement techniques and selecting what sort of information is most useful. Some mathematical ideas such as principal components analysis and entropy will be briefly introduced. Since this text is primarily aimed at chemists rather than statisticians, these concepts will not be discussed in detail: this is left to later chapters and other texts.

4.2 AN INTRODUCTION TO PRINCIPAL COMPONENTS ANALYSIS

At this point it is important to understand one of the basic methods used in chemometrics, namely, *principal components analysis*. We will attempt to explain the basis of such an approach without the need for detailed matrix algebra or mathematics. Further sophistications and refinements of this method will be discussed in Chapters 6 and 7.

Consider a chemical experiment consisting of recording the NMR spectrum of a mixture of three different compounds. Each compound might give rise to several peaks, so the chemist is faced with interpreting a spectrum of, for example, 20 peaks. He might then record several spectra with each of these compounds present in different absolute amounts. From the data, consisting of N samples (or spectra) and I (=20 in this example) variables (or peaks), the chemist is interested in N samples but only really P (=3 in this example) *reduced variables* where $P \leq I$. One way of looking at the problem is to imagine each spectrum represented as a point in I–dimensional space, defined by the axes x_1, ... , x_I where each axis represents the intensity of each of the I peaks. But each spectrum could be also represented as a point in P dimensional space, each dimension being the absolute amount of each of P compounds. The P dimensional problem is easier to handle because it involves less numbers and if

P is small (e.g. 3) is easier to visualize graphically. Principal components analysis is a form of *dimensionality reduction* .[1]

Returning to the problem of choosing and comparing analytical methods, we noted above that we rarely measure chemical information directly. So, for example, a concentration may be measured by recording a voltage or the volume of solvent in a burette. Similarly, certain types of chemical information may be measured by a series of observations or a *multivariate* dataset. Within a dataset of 20 measurements (e.g. 20 GC or pyrolysis mass spectral (MS) peaks) there may be three or four interesting chemical trends. The chemist is not really interested in raw data (in the univariate case, for example, he is not directly interested in voltages but in concentrations) but in parameters calculated from these data. The chemist might want to compare various measurement techniques and an important question to ask is what sort of information is obtained from each technique. For example, do pyrolysis ms and GC provide similar chemical information for a particular problem or not? By performing dimensionality reduction on the data obtained from each of these techniques the chemist can then compare the information obtained.

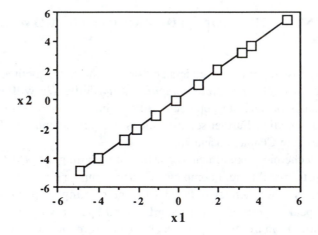

Fig. 4.1 - Two perfectly correlated variables.

A very simple case of dimensionality reduction is when two parameters, x_1 and x_2, are recorded. Let us assume that $x_1 = x_2$ (one parameter may be the value of the concentration of a compound estimated using one wavelength in uv /

[1] In fact the problem posed above is normally solved by factor analysis methods (see Chapter 6), but the basis of both approaches is similar.

visible spectroscopy, and the other parameter may be the concentration monitored at another wavelength). The graph is illustrated in Fig. 4.1. Instead of quoting two parameters all the time, it is usual to cite only one parameter. Although the data have been recorded in two dimensions, in fact they are only one dimensional in nature since all the points lie on a straight line. A simple way of reducing the dimensionality from two to one is to project all the points onto the line $x_1 = x_2$ and replace the two dimensional graph by a one dimensional graph (Fig. 4.2). Instead of citing both values of x_1 and x_2 we cite only one value, namely

$$x_1' \quad = \quad \sqrt{x_1^2 + x_2^2} \qquad\qquad (4.1)$$

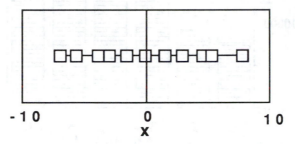

$$-1\,0 \qquad\qquad 0 \qquad\qquad 1\,0$$
$$X$$

Fig. 4.2 - Projection of data from Fig. 4.1 onto one dimension.

which happens to equal $\sqrt{2}\, x_1$ in this particular case. This is the *distance* of each point from the origin along the new axis. We have reduced the dimensionality of the data (so made it easier to handle) without losing any essential information.

One conceptual difficulty with this analysis is that we have chosen to use $\sqrt{2}$ x_1 rather than, for example, x_1 or $(x_1 + x_2)$. Why is this? The choice of distance relates to the crucial concept of *scaling*. Chemists go to great length to physically scale variables. If a concentration is estimated by recording a voltage, we do not cite a voltage in papers, but a concentration, and this concentration is used in further calculations. An important feature of chemometrics is that it helps us look at the relative importance of trends. Consider the histogram in Fig. 4.3, which might be of the long term sales figures of a product. It looks as if the sales of the product are increasing dramatically. However, the scale of the axis is between 100,000 and 120,000 per month. Change the axes to 0 and 500,000 and we see that the sales of the product hardly changes (Fig. 4.4). Let us now try to compare the sales with the production figures. When we scale the production figures between 0 and 500,000 the histogram (Fig. 4.5) *appears* identical to Fig. 4.3, so we might conclude that production and sales are obeying identical trends and everything is going well. In fact, we should compare Fig.

4.5 to Fig. 4.4 and we see that there are likely to be severe economic problems since production severely oversteps sales. The chemist decides, himself, what scale he wants to use. This decision is made after a lot of thought and it would be wrong for a computer program or method of calculation to distort these scales. Fig. 4.2 is merely a one dimensional projection through two dimensional space and leaves scaling (*geometric distance*) unchanged.

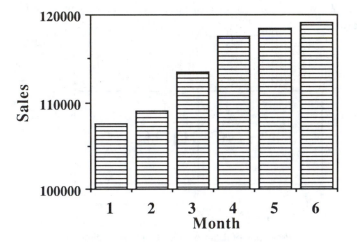

Fig. 4.3 - Monthly sales figures of a product.

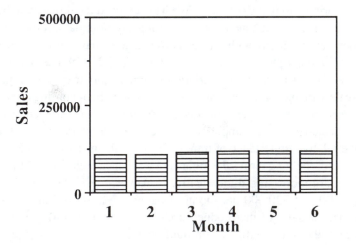

Fig. 4.4 - Data of Fig. 4.3 plotted on a different scale.

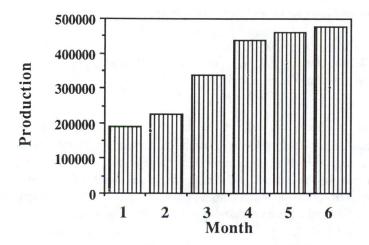

Fig. 4.5 - Monthly production figures of the product.

This concept of scaling is so important in chemometrics that it is probably best to consider in more detail exactly why we might expect the value of $\sqrt{2}$ to appear in the case discussed above. Let us assume that x_1 and x_2 are spectroscopic peaks of roughly similar mean intensity. As discussed in earlier chapters, *noise* is generally imposed upon signals. There are many possible ways of modelling noise distributions, some of which will be discussed in Chapter 5, but let us, for simplicity, employ a *uniform* distribution where noise can take any, equally likely, value between N and $-N$, with, therefore, a mean of 0. The interested reader can derive the expressions below for more complex noise distributions.

The *signal to noise ratio* is normally defined as the ratio of the intensity of a signal to the root mean square noise. If the intensity of a signal is given by s, then the mean square noise for our uniform distribution will be given by

$$r_1 = \int_{-N}^{+N} v^2 dv = \frac{2N^3}{3} \tag{4.2}$$

where v is the value of the noise at any given point. For two signals, this expression becomes

$$r_2 = \frac{1}{2} \int_{-N}^{+N} \int_{-N}^{+N} (v_1^2 + v_2^2) dv_1 dv_2 = \frac{4N^3}{3} \tag{4.3}$$

But, the signal strength doubles, so if it is s for one peak, it becomes $2s$ for two peaks. Therefore the signal to noise ratio for one peak is

$$sn_1 \quad = \quad \frac{s}{\sqrt{2N^3/3}} \tag{4.4}$$

and for two peaks it is

$$sn_2 \quad = \quad \frac{2s}{\sqrt{4N^3/3}} \tag{4.5}$$

giving

$$sn_2 \quad = \quad \sqrt{2}\ sn_1 \tag{4.6}$$

Therefore two peaks yield $\sqrt{2}$ more information than one peak (this makes a number of assumptions about the noise, but most symmetric noise distributions have this property). Now, consider the case where we monitor three peaks, x_1 and x_2 arising from compound A and x_3 from compound B. We might be interested in whether there is a relationship between the occurrence of compound A and compound B. What we really want to do is to see whether the concentration of A (as estimated from x_1 and x_2) *correlates* to the concentration of B (as estimated from x_3). But we know more about the concentration of compound A as we have taken more measurements on A and so we want to use this information in our calculations and not throw this away since we have taken the time and trouble to record the second peak due to A. So we *weight* the data for A as a little more significant to the data for B. It is sensible to weight these two datasets by the relative signal to noise ratios.

Naturally we might choose to ignore or even not to measure x_2. We might know that one of these measurements is better (more precise) than the other. Obviously we can utilize this information, but this must be under the control of the chemist, and not implicit in a computer program; he should be able to specify the significance of the variables how he wishes, by weighting the variables, or converting to different scales, which, geometrically, is equivalent to expanding, contracting, translating (and even rotating as discussed below) the axes. An example of the influence of scaling on interpretation of the data is illustrated when comparing Figs 4.3 and 4.4. The mathematical process of dimensionality reduction, though, must keep the geometry the same, otherwise all sorts of problems might occur. So principal components analysis and most (not all) methods for dimensionality reduction in chemistry maintain geometric (or Euclidean) distances. It of vital importance to understand this and also to appreciate that the chemometrician himself has control over the scales and significance of variables and should utilise any knowledge of the system he may have to sensibly scale axes if so desired, prior to principal components analysis.

We have discussed above only a very simple case of dimensionality reduction. Typically we might be interested in reducing 20 variables to three, i.e. in finding three dimensional cross-sections in 20 dimensional space, so the problem is slightly more complex.

Consider now three variables, x_1, x_2 and x_3 such that the first two variables can take on any value, and the third is given by

$$x_3 \quad = \quad 2x_1 + x_2 \tag{4.7}$$

It is clear that, although three measurements have been taken, the dataset is only two dimensional. How do we select these two dimensions? It is, of course, possible to select any two variables, x_i and x_j, and discard the third. Although this retains all the information, it is not a principal component solution. The reason why it is not is that the geometry is distorted. Eq. (4.7) is the equation for a two dimensional cross section in three dimensional space, in other words, a plane. In the example above, we *projected* two dimensional data onto a straight line. Analogously we want to project the three dimensional data onto a plane. In fact if we find the right plane, all the data lie in this plane anyway.

Chemists, however, are not always too interested in equations for planes. They are interested in interpreting data, and what we really want to do is replace the three values of x_1, x_2 and x_3 by two new values x_1' and x_2', without distorting the space. In fact another condition is met by PCA in that all axes are *orthogonal* to each other, that is at right angles to each other. So the x_1' and x_2' axes are chosen to be orthogonal. Thus PCA becomes an exercise in geometry.

One solution is

$$x_1' \quad = \quad (1/\sqrt{5})x_1 + (2/\sqrt{5})x_3 \tag{4.8}$$

$$x_2' \quad = \quad (\sqrt{6/5})x_2 \tag{4.9}$$

Simple matrix algebra is described in the appendix, but it is easy to verify this solution, as follows. The magnitude must remain the same, so that,

$$x_1'^2 + x_2'^2 \quad = \quad x_1^2 + x_2^2 + x_3^2 \tag{4.10}$$

as can readily be seen from Eqs (4.7), (4.8) and (4.9). In addition the coefficients α_{ij} given by

$$x_i' \quad = \quad \sum_{j=1}^{3} \alpha_{ij} x_j \tag{4.11}$$

should form orthogonal vectors, so, in vector notation

$$\begin{pmatrix} \alpha_{11} \\ \alpha_{12} \\ \alpha_{13} \end{pmatrix} \cdot \begin{pmatrix} \alpha_{21} \\ \alpha_{22} \\ \alpha_{23} \end{pmatrix} \quad = \quad 0 \tag{4.12}$$

in the example above.

It is usual to normalize the coefficients so that

$$\sum_{j=1}^{3} \alpha_{ij}^2 \quad = \quad 1 \tag{4.13}$$

for each new variable. Although this is obviously true for Eq. (4.8), it is not immediately obvious in the case of Eq. (4.9). In fact, if we use Eq. (4.7), we can modify the Eq. (4.9) so that

$$x_2' \quad = \quad -2/\sqrt{30}\, x_1 + \sqrt{(5/6)}\, x_2 + 1/\sqrt{30}\, x_3 \tag{4.14}$$

For data related by Eq. (4.7), this new definition of x_2' is identical to the definition of Eq. (4.9), and we will use Eq. (4.14) in preference to Eq. (4.9).

But there are many possible solutions to this problem that satisfy Eqs (4.10) and (4.12). These, geometrically, can be visualized as rotation in the plane given by Eq. (4.7). One way of generate these solutions is to compute values x_i'' given by

$$x_1'' \quad = \quad (\cos \theta)\, x_1' + (\sin \theta)\, x_2' \qquad\qquad (4.15)$$
$$x_2'' \quad = \quad -(\sin \theta)\, x_1' + (\cos \theta)\, x_2' \qquad\qquad (4.16)$$

where θ is an angle of rotation, as can be shown by simple geometry.

This multitude of solutions has the possibility of disastrous implications. A chemist might may want to put three NMR intensities into a computer program and emerge with two reduced principal components or reduced variables. However, if there are an infinite number of solutions then the chemist can obtain an infinite number of answers, and two computer programs might provide him with completely different answers on the same data. Clearly some further thought must be given to this problem.

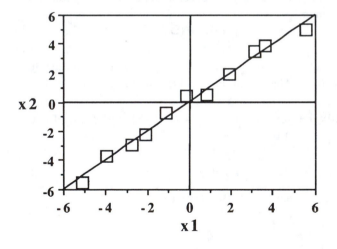

Fig. 4.6 - Two variables that are probably correlated but contain some experimental error.

A clue to the difficulty comes from reconsidering the problem when $x_1 = x_2$. In many texts (as here) principal components analysis (PCA) is introduced as a method for projecting two dimensional data onto a straight line. Such explanations omit some vital aspects of PCA as we are beginning to see. Normally noise is imposed upon the data, so x_1 and x_2 are not exactly equal. We discuss the problem of experimental errors in all other chapters, and this chapter is no exception. In practice,

$$x_{in} \;=\; \hat{x}_{in} + \eta_{in} \qquad\qquad (4.17)$$

where η_{in} is a noise function and x_{in} now stands for the nth observation of x_i, and $\hat{x}_{in}$ the corresponding predicted response. Note we have now slightly redefined our problem. Instead of dealing with pure functions, we are dealing with samples as well. In chemometric terms we have measured i variables (two in the case in question) for n samples. Each measurement comes with its own experimental error. The observations now no longer exactly fit a straight line. An example is illustrated in Fig. 4.6. However, if we still suspect the data is one dimensional, we still want to represent these data by a single principal component. In fact this process of dimensionality reduction is very valuable because it helps reduce the experimental noise and can be looked at as a smoothing operation.

The normal approach is to draw a best fit straight line through the data. This is defined as the straight line that minimizes the sum of square errors between the observed and expected points. Superficially PCA is similar to the methods for model building discussed in Chapter 2, which lead to ANOVA, but there is one crucial difference. In Chapter 2 it is assumed that there are no errors in the independent (or x) variable and all the errors are in the response. Thus, if x_1 is a factor and x_2 a response, the errors for linear regression and consequent response modelling as discussed in Chapter 2 are given by

$$S \;=\; \sum_{n=1}^{n=N} (\hat{x}_{2n} - x_{2n})^2 \qquad\qquad (4.18)$$

using the terminology above. For a linear model, of the form $x_2 = x_1$, as discussed here, the error simply reduces to

$$S \;=\; \sum_{n=1}^{n=N} (x_{1n} - x_{2n})^2 \qquad\qquad (4.19)$$

since the estimated value of x_2 is x_1. However, in the case of principal components analysis there are likely to be similar errors along both the axes. If one axis represents one peak in an NMR spectrum and the second axis another peak, why should one peak be error free and the other contain errors? So in PCA we compute the *orthogonal sum of squares,* that is the projected distance between the observed points and the predicted *model*. The two forms of estimating errors (normal linear type linear regression and principal components analysis) are illustrated in Fig. 4.7. This distance is the minimum distance between the predicted and observed points.

If the straight line is at an angle θ from the origin (i.e. 45° for $x_1 = x_2$) then for a pair of points x_1 and x_2, we can show by simple geometry

$$x_1' \;=\; (\cos\theta)\, x_1 + (\sin\theta)\, x_2 \qquad\qquad (4.20)$$

$$x_2' \quad = \quad -(\sin \theta) x_1 + (\cos \theta) x_2 \qquad (4.21)$$

Ideally, if there are no errors, x_2' would equal 0, but, in practice it is the orthogonal distance of the point from the new principal component axis. Therefore the overall error will be given by

$$S = \sum_{n=1}^{n=N} x_2'^2 = \sum_{n=1}^{n=N} ((\sin^2\theta)x_{1n}^2 + (\cos^2\theta)x_{2n}^2 - 2(\cos\theta.\sin\theta)x_{1n}x_{2n}) \quad (4.22)$$

which, since $\theta = 45^\circ$ in our example, reduces to

$$S \quad = \quad 0.5 \sum_{n=1}^{n=N}(x_{1n} - x_{2n})^2 \qquad (4.23)$$

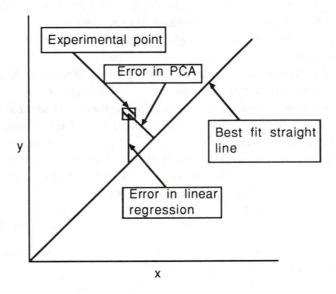

Fig. 4.7 - Different methods of estimating errors: linear regression and principal components analysis.

Note that the root mean residual sum of squares for a PCA calculation is now $1/\sqrt{2}$ of that for a normal linear regression calculation in the case cited (the value of the constant is dependent on θ and $\cos 45^\circ = 1/\sqrt{2}$). We have already discussed why this factor appears. Obviously the ratio of the sum of squares in normal multilinear regression (as used to fit response surfaces, for example) and PCA type regression will depend on the value of θ and so the model. For a two factor example, this difference is relatively trivial, but for several factor examples, it can make a considerable difference. Effectively in normal regression we make inherent assumptions as to the nature of the errors. For example if we are fitting a single (univariate) response to a two factor experiment, we assume that only one of the three variables (the response) is subject to error. A three variable PCA calculation will assume all three variables are equally subject to error. As commented in Chapter 2, relatively little work has been performed on multivariate experimental design in which both response and factors are multivariate. If, however, we designed an experiment with three factors and three responses we could place each experiment in six dimensional space, but use regression methods that assume that there are only errors in three of the dimensions.

Because PCA adopts a rather more complicated approach to errors than ANOVA, PCA models are usually always linear rarely include higher order terms.[1] Data are also normally mean centred before PCA calculations, that is

$$x'_{in} = x_{in} - \bar{x}_i \qquad (4.24)$$

where there are N samples, 1 to n, and $\bar{x}_i$ is the mean of variable i. In the numerical example we give below all data are already mean centred, that is the data all have means of 0. Thus the result of a PCA calculation that reduces a dataset of I variables to one of P variables where $P < I$ will be to find x''_1, ... x''_P so that

$$x''_{in} = \sum_{j=1}^{j=I} \alpha_j x'_{jn} \qquad (4.25)$$

Let us return to the data of Fig. 4.6. The PCA projection onto one dimension is illustrated in Fig. 4.8. In the example above, we discussed the problem of reducing dimensionality from three to two. Instead of the axes being x_1 and x_2 they could be the axes given by Eq. (4.8) and Eq. (4.14). We commented that, without further information there would be an infinite number of solutions of the forms given by Eq. (4.15) and Eq. (4.16). However, once projected onto the plane the data will form a pattern on the plane. The data in Fig. 4.6 could equally well be a two dimensional projection of a three dimensional dataset. We employ exactly the same principles for finding the principal components, i.e. we find the first principal component namely the axis that minimizes the orthogonal sum of squares, which will in this case be (using Eqs (4.8) and (4.9) and remembering that $\cos 45° = 1/\sqrt{2}$)

$$\begin{aligned} x''_1 &= (1/\sqrt{2})\,[(1/\sqrt{5})x_1 + (2/\sqrt{5})x_3\,] + (1/\sqrt{2})\,[-2/\sqrt{30}x_1 + \sqrt{(5/6)}x_2 + \sqrt{(1/30)}x_3] \\ &= 0.058x_1 + 0.645x_2 + 0.762x_3 \qquad (4.26) \end{aligned}$$

The axis orthogonal to this is the second principal component, given, in this case, by

$$x''_2 = -(1/\sqrt{2})\,[(1/\sqrt{5})x_1 + (2/\sqrt{5})x_3] + (1/\sqrt{2})\,[-2/\sqrt{30}x_1 + \sqrt{(5/6)}x_2 + \sqrt{(1/30)}x_3]$$

[1]Higher order PCA has been proposed for use in psychometrics. The treatment of errors is rather obscure, but higher order terms of the form x_1^2 etc. are added as extra axes. It has been suggested that this approach could be applied in chemistry but as yet there is very little software available and no clear physical explanation as to how this approach relates to experimental errors.

$$= \quad -0.574x_1 + 0.645x_2 - 0.503x_3 \tag{4.27}$$

For readers who use standard linear regression software, it is important to recognize that the PCA "best fit" straight line is not the same as the answer that comes from simple linear regression, because of the different nature of the errors. It is easy to show that the angle this straight line makes with the origin is given by

$$\tan 2\theta \quad = \quad 2\sum_{n=1}^{n=N} x_{1n} x_{2n} / \left(\sum_{n=1}^{n=N} x_{1n}^2 - \sum_{n=1}^{n=N} x_{2n}^2 \right) \tag{4.28}$$

for two factors (see appendix). It is possible to derive analytical solutions for more than two factors, but the algebra is extremely complex and people normally prefer iterative computational algorithms.

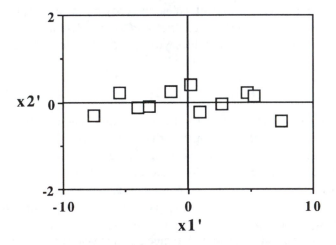

Fig. 4.8 - Projection of data from Fig. 4.6 onto best fit straight line, using principal components criteria.

It is constructive to demonstrate this by a simple calculation. Table 4.1 is a raw dataset, with properties given by Eq. 4.7 (each variable is already mean centred). From preliminary inspection it looks roughly as if all three variables show approximately the same trend, but PCA will help us sort out and quantify this trend. The transformation of Eqs (4.8) and (4.14) is tabulated in Table 4.2. Finally the PCA solution is given in Table 4.3. The variables x_1'', x_2'' and x_3'' are the three principal components. The value of the third principal component is exactly zero as all points can be projected exactly onto a plane. However the second principal component includes some information. The amount of information obtained from each principal component can be assessed by using sums of squares in the same way as in ANOVA analysis described in Chapter 2.

The total sum of squares for all the data remains constant at 228.24 no matter how the data is transformed, giving us confidence that geometry and distances remain the same (Table 4.4).

Table 4.1 - Raw data for example of Tables 4.2 to 4.5

Expt	x_1	x_2	x_3
1	−0.214	0.409	−0.019
2	0.184	0.501	0.869
3	−0.224	−0.715	−1.163
4	0.188	1.733	2.109
5	−0.119	−2.035	−2.272
6	0.140	3.185	3.464
7	−0.161	−2.665	−2.987
8	0.220	3.516	3.955
9	−0.439	−3.372	−4.249
10	0.685	4.511	5.881
11	−0.259	−5.069	−5.587

Once the principal components have been computed we see that the majority of variability lies along the first principal component. In fact 99.69% of the sum of squares is accounted for by the first principal component and only 0.31% by the second principal component. If we did not know the trends in the data in advance, we would probably conclude that there is only one significant principal component and the second represents such a small amount of variability that it is effectively purely noise. In fact tests for the number of significant principal components in a dataset are of considerable chemometric interest. Principal components are normally ranked in descending order of their sum of squares, and various tests are used to determine the true dimensionality of the data. These are discussed in further detail in Section 6.2. Chemometricians must not get confused by the ready availability of user friendly chemometrics software. Each software author has his or her favourite methods for testing the significance of each factor and often builds these approaches into the programs. It is possible to perform extensive computations on large numbers of datasets and compare the number of principal components computed using various criteria. Often different approaches for assessing the number of significant components provide different answers. Some investigators claim that the nature of the noise distribution

Table 4.2 - Data of Table 4.1 transformed according to Eqs (4.8) and (4.9)

Expt	x_1'	x_2'
1	−0.113	0.448
2	0.860	0.549
3	−1.141	−0.783
4	1.970	1.898
5	−2.085	−2.229
6	3.161	3.489
7	−2.744	−2.919
8	3.636	3.852
9	−3.997	−3.694
10	5.566	4.942
11	−5.113	−5.553

Table 4.3 - Principal components of data in Table 4.1

Expt	x_1''	x_2''	x_3''
1	0.237	0.397	0
2	0.996	−0.220	0
3	−1.360	0.253	0
4	2.735	−0.051	0
5	−3.050	−0.102	0
6	4.702	0.232	0
7	−4.004	−0.124	0
8	5.295	0.153	0
9	−5.438	0.214	0
10	7.430	−0.441	0
11	−7.542	−0.311	0

determines the effectiveness of various approaches for determining the significant number of components. Indeed any extra information about a system

Table 4.4 - Sums of squares for variables in Tables 4.1 to 4.3

Table	Parameter	Sum of squares
4.1	x_1	1.00
	x_2	95.10
	x_3	132.14
4.2	x_1'	114.12
	x_2'	114.12
4.3	x_1''	227.53
	x_2''	0.71
	x_3''	0.00

Table 4.5 - Standardized principal components from Table 4.3

Expt	1st PC	2nd PC
1	0.050	1.486
2	0.209	−0.824
3	−0.285	0.948
4	0.573	−0.191
5	−0.640	−0.381
6	0.986	0.869
7	−0.839	−0.464
8	1.110	0.572
9	−1.140	0.803
10	1.558	−1.653
11	−1.581	−1.166

is always useful. If it is known that the measurement noise obeys a Gaussian distribution, for example, this information should be put to use: instead of simulating or computing vast numbers of PCA calculations in the presence of various models of noise, it is often possible to use simple mathematical arguments for comparing the effectiveness of various criteria in determining the number of significant components.

Generally the output of PCA package is a graph of what are called "scores" (equivalent to the variables x_i'') or "loadings" (equivalent to the coefficients α_i) plots. The use of these and related graphs will be described in more detail in Chapters 6 and 7. Extreme care should be taken when interpreting the scales of these graphs. Often chemometricians standardize the scales, so that each principal component has zero mean and unit variance. Because each principal component always has zero mean, this latter transformation does not influence the principal component. However, dividing by the standard deviation can make a large difference to the relative scales. The standard deviation of x_1'' is 4.54, whereas for x_2'' it is 0.25. These scaled principal components are tabulated (Table 4.5). Often this and related scaling is performed automatically by chemometrics software, yet is frequently misleading. The motivation is good graphics rather than good statistics. Another common transformation is to reflect the principal components in the origin so that one package might give exactly negative values to the other package. This is because $x = \sqrt{(\pm x)^2}$ and there are no generally accepted conventions. It is vital, though, that the user of chemometrics software is extremely cautious in interpretation of the scales on principal components and related graphs, and does not get carried away by elaborate graphical output.

The example given in Table 4.1 is a relatively simple one. More typical chemical applications might involve 20 measurements (e.g. 20 GC peak intensities) which are reduced to three or four significant principal components. Under such circumstances, the main trends in the data are not always obvious at first glance, and it is necessary to use one of many computer packages available for PCA. However the principles are the same, in that orthogonal axes that explain successively decreasing variability (assessed by the sums of squares) in the data are computed. The first few components are then taken as a hopefully noise free model of the data.

Many computational algorithms have been developed for PCA, but a detailed description is outside the scope of this text.[1] Some methods compute all components at once, others find the most significant component first (the axis of most variability) and then the next component (the most significant axis orthogonal to the first) and so on. The mathematics of PCA can be described by matrix algebra (see appendix) but it is not really necessary to understand this unless one is, in turn, trying to write a PCA program. It is, however, vital to appreciate the importance of scaling, assessment of errors, criteria for the number of components, the role of noise and so on, and it is important to have these crucial concepts to mind when using PCA software.

---

[1]The NIPALS algorithm is described in the appendix.

4.3 MULTIVARIATE APPROACHES FOR COMPARING MEASUREMENT TECHNIQUES

The theme of this chapter is how best to choose and optimize analytical conditions. One of the commonest problems to face a chemist is which, of a multitude of available methods, should he or she employ in order to obtain information about a sample? Should he or she use NMR, or pyrolysis, or HPLC, or GC, or MS or IR (infrared) or what? Ideally, perhaps, he should employ *all* these methods, but time is frequently limited. By employing a few selected measurements, can he or she obtain similar information?

Let us return to the example of Table 4.1. PCA suggests that all three measured variables provide roughly similar information. Is it possible to get away with measuring only two variables, and if so, which two variables are most useful? Let us compare the information content of variables x_1 and x_2. A first step is to plot a graph of x_1 against x_2 (Fig. 4.9). This graph suggests a roughly linear trend. The least squares best fit linear equation is given by

$$x_1 \quad = \quad 0.087 \, x_2 \qquad\qquad (4.29)$$

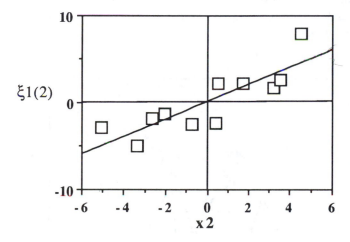

Fig. 4.9 - Graph of $\xi_{1(2)}$ against x_2 as described in the text.

assuming all errors are in x_1. We can see, however, that there is some scatter about the straight line. One approach might be to *scale* x_1 so that there is maximum overlap with x_2. Let us call this new variable $\xi_{1(2)}$. It is easy to compute this value, since

$$\xi_{1(2)} \;=\; \frac{x_1}{0.087} \tag{4.30}$$

This variable, together with the corresponding variable $\xi_{3(2)}$ is tabulated in Table 4.6. If two variables are perfectly matched, these two parameters, together with x_2, should be identical. All that we have done is scaled and changed the intercept of the variables in order to obtain maximal overlap. This is equivalent to taking three measurements from a spectrum and seeing whether these measurements, which may be of different relative intensities (hence the scaling), are really fairly similar.

Fig. 4.10 shows the correspondence between x_2 and $\xi_{1(2)}$. If both parameters contained the same information we would expect the two sets of measurements to overlap exactly. This is clearly not the case and there are quite a few crossed lines. How can we assess how similar these measurements are? Again the sum of squares comes to our rescue. It is merely necessary to calculate

$$S \;=\; \sum_{n=1}^{n=N} (x_2 - \xi_{j(2)n})^2 \tag{4.31}$$

when comparing variable j to variable 2, for N observations in a dataset. We find that for x_1, this sum is 37.68 but for x_3 it is only 1.43 on the same scale. Therefore x_2 and x_3 measure roughly similar information, but x_1 measures different information.

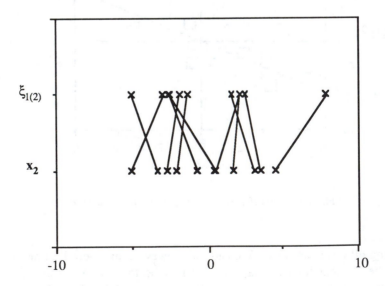

Fig. 4.10 - Comparison of x_1 and x_2 as desribed in the text.

Table 4.6 - Values of the scaled best fit parameters when x_1 and x_3 are scaled
to fit x_2 as closely as possible as described in the text

Expt	x_2	$\xi_{1(2)}$	$\xi_{3(2)}$
1	0.409	−2.466	0.350
2	0.501	2.120	0.429
3	−0.715	−2.584	−0.612
4	1.733	2.164	1.484
5	−2.035	−1.365	−1.742
6	3.185	1.608	2.727
7	−2.665	−1.857	−2.282
8	3.516	2.528	3.011
9	−3.372	−5.051	−2.887
10	4.511	7.884	3.863
11	−5.069	−2.982	−4.341

   This type of test can be used to answer more sophisticated questions. Is the information obtained from variables x_2 and x_3 equivalent to the information from variable x_1 alone? The first variable may be an NMR measurement and the second and third may be mass spectrometric measurements. If this indeed is so, it is only necessary to employ one of these two techniques and we can save time. The way this calculation could be performed is to compute the first principal component for x_2 and x_3 and compare this with x_1.

   This procedure can be extended to more complex situations and is called *procrustes analysis*. Instead of comparing three variables, why not compare three measurement techniques? Each variable might then represent the first principal component from each measurement technique. For example we might record the mass spectra, NMR spectra and gas chromatograms for a series of samples. The data for each technique is reduced to one principal component and these three principal components are compared. We can then directly decide which techniques contain similar and which new information.

   More commonly, though, a given measurement method yields several significant principal components and it is possible to put these to use, simply by comparing samples in P dimensional space, where P is the number of significant components. We have more options, though, than merely scaling and shifting the origin in higher dimensional space, and it is usual to rotate and reflect measurements until they best overlap (minimum sum of squares). An example of three measurements (a triangle) in two dimensional space is given in Fig. 4.11. So a general procedure for comparing measurement technique B to A is as follows.

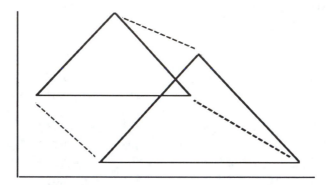

Fig. 4.11 - Procrustes analysis in two dimensions: the larger triangle is contracted and translated to overlap with the smaller triangle. The corners of each triangle represent three measurements; each triangle represents a different measurement technique. For more complex situations, rotations are also used to maximize overlap.

(1) Compute the principal components of A and keep the P most significant components.
(2) Compute the first P components arising from technique B (note that we need an equal number of components from each technique so that the principal component space is of the same order).
(3) Scale (keeping the *relative* dimensions of each axis the same), translate, rotate and reflect the observations on B so that there is maximal overlap (minimal sum of squares distance) between A and B. The configuration where A and B overlap most is called the *consensus configuration*. Note that in the examples above each variable has already been mean-centred, so translation is not necessary. Rotation is important in multidimensional contexts.
(4) Calculate the minimum sum of squares distance and decide whether there is any significant difference between A and B.

There are a number of algorithms for procrustes analysis but the details need not concern us here. However, in some cases it is possible to compare measurements with a different number of principal components (step 2). If technique A has P significant principal components and technique B has $P-i$, then principal components $P-i+1$ to P for technique B are set to 0. Also several criteria of significance can be used in step 4. It is probably best to examine each problem on its merits and not try to automate the criteria.

The method of procrustes analysis for comparing analytical techniques can be extended to a large variety of situations. For example, if we routinely employ three methods (e.g. AA, HPLC and pyrolysis) for the analysis of sets of samples (e.g. routine analysis of river water for monitoring pollution) does a new approach (e.g. MS) provide any more information? In such a case it might be sensible to compare the principal components of the combined AA, HPLC and pyrolysis measurements to the principal components of the MS measurements. Another example is if 50 tests are used (e.g. clinical trials) each providing one piece of information. Can we reduce these to 20 tests? We could compare the information from 30 tests with that from 20 tests. Obviously the software becomes quite complex here if we want to compute information content from any possible combination of 20 tests, or work out the minimum number of tests necessary, and if such a situation is reached, it is important to balance the time required to develop software against the chance of not quite reaching an optimal solution. Probably a great deal depends whether the particular laboratory has programmers or technicians on its payroll.

It is, of course, not necessary to use principal components reduction as a first step in procrustes analysis: any sets of data of the same dimensionality can be compared by this technique. In fact, several of the steps of procrustes analysis become redundant if applied to data reduced by PCA: for example, there is no need to translate two datasets as their mean will always be zero. Any other form of dimensionality reduction such as those described in Chapters 6 and 7 can also be employed.

There are also other general multivariate approaches besides procrustes analysis to this problem, including methods such as partial least squares and canonical variates. These approaches will be discussed, in other contexts, in later chapters.

4.4 INFORMATION THEORETICAL APPROACHES TO COMPARING ANALYTICAL METHODS

A radically different approach for comparing analytical methods arises from information theory. Below we discuss simple concepts of *entropy* and *bits* of information, which will be expanded upon in Section 5.4.

Consider performing a simple experiment to classify a sample into one of two classes or states. The answer from the experiment may be that a piece of meat comes from a cow raised by one farming method or by another. Let us, initially, assume that each answer is equally likely. A histogram of probabilities is illustrated in Fig. 4.12. If the experiment provides a yes / no answer and is effective, no further work is required. The experimental results could be coded so that one answer is given by "1" and another by "0". This is equivalent to expressing the experimental result as a single binary digit or a bit. Now, consider a more complex problem, that of distinguishing 16 different states: for example these might correspond to 16 different stereoisomers of a compound. Only four experiments that give yes / no answers are required since $16 = 2^4$ and so the experimental results could be coded as a binary number between 0000 and 1111. We can imagine the desired results for each isomer expressed as 4 bits (= binary digits) of information. However the experiments need to be *designed* properly. It would be no use if each experiment yielded identical information. Information theory can help us choose experiments that provide, overall, the correct amount of information as described below.

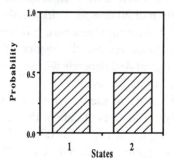

Fig. 4.12 - Two equally probable states.

If the four tests are not designed properly two or more isomers might yield identical results. The way to rectify the situation is either to increase the number of tests or change the nature of the tests themselves. We need some way of determining whether we have performed sufficient tests to uniquely distinguish the 16 isomers, and also to compare the efficiency of two or more sets of tests.

We can extend the arguments above to more general discrete distributions. If we want to distinguish N equally probably discrete states, then the desired amount of information is given by

$$I \quad = \quad \log_2 N \tag{4.32}$$

where I is the defined as the number of bits of information. To distinguish 16 states, we need 4 bits of information, as discussed above. However, states are not always equally likely. If each of the states has a probability of p_n then the desired information content is given by

$$S \quad = \quad -\sum_{n=1}^{n=N} p_n \log_2 p_n \tag{4.33}$$

which reduces to Eq. (4.29) when $p_n = 1/N$ for all N states. The expression on the righthand side of Eq. (4.30) is often called the *Shannon entropy* or simply entropy (there are other definitions of entropy as discussed in Section 5.4, but we will restrict ourselves to this definition at present). We will define entropy by S in the remainder of this section but the reader should remember that entropy and information content are effectively the same. Fig. 4.13 illustrates the case of unequal probabilities for four states, and the data are tabulated in Table 4.7. A simple calculation shows that the entropy is 1.818.[1] This entropy is less than the entropy for four equally likely states which is exactly 2 bits. That means it is easier to distinguish if probabilities are known to be unequal. This makes intuitive sense. If we knew that the probabilities of two of the four states were 0.5 and of the other two states were 0 then the problem would reduce to one of distinguishing two states rather than four, and so is easier.

Of course, most distributions encountered by chemists are continuous distributions and so Eq. (4.29) can be generalized still further to give

$$S \quad = \quad - \int_{x_{lo}}^{x_{hi}} p(x)\log_2(p(x))dx \qquad (4.34)$$

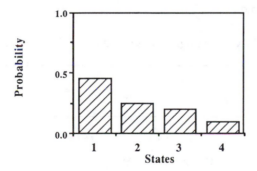

Fig. 4.13 - Probabilities of states listed in Table 4.7. This histogram could represent a simple chromatogram or spectrum sampled at four points, and, as discussed in the text, has an entropy of 1.818.

where x is a variable being observed and x_{lo} and x_{hi} are the limits for the variable (for example the proportion of a solvent in a mixture may vary between 0 and 1). One important point to note is that the more that is known about a system (e.g. the probabilities of taking on different values), the more certain we can be of the results of the experiment. If nothing is known about a system, normally a flat distribution is assumed, with every state having equal probability. This is the

[1]In this chapter we will use logarithms to the base 2 so that entropy and bits of information are numerically identical.

distribution of *maximum entropy*. The more the information that is known in advance the lower the entropy. A system with entropy 0 is one in which

Table 4.7 - Probabilities of states illustrated in Fig. 4.13

State	Probability
A	0.45
B	0.25
C	0.2
D	0.1

everything is known prior to experimentation. If certain knowledge of a system is not incorporated in the model of Eq. (4.30) then this may result in unnecessary experimentation. Hence model building prior to experimentation is an important aspect of properly using information theoretical methods for choice of measurement technique.

Table 4.8 - Chromatographic separation of four isomers

Condition	A	B	C	D	E
Isomer					
1	X	X	Y	X	Y
2	X	Y	X	X	X
3	Y	X	Y	X	X
4	Y	Y	X	Y	X

Note : The X and Y stand for unresolved clusters of peaks, so, for example, using condition D, isomers 1, 2 and 3 coelute, whereas isomer 4 elutes separately. For each condition in this example, there are only two clusters of peaks.

Let us return to the discrete case. Consider choosing analytical conditions for the separation of four, equally likely, isomers chromatographically. Exactly 2 bits of information are required from the experiment. Let us imagine we choose a number of chromatographic conditions, and each method separates the isomers into two unresolved clusters of peaks, X and Y. In order to work out the identity of an unknown compound, we try out a condition, and see which cluster of peaks (which may be characterized by a retention time or R_f value, for example) the compound belongs to. We want to find out which conditions are most informative and which combination of conditions uniquely characterizes each isomer. Five different conditions are tabulated in Table 4.8. The entropy of each condition can easily be computed. For condition E, the probability of obtaining an isomer that elutes at position Y is 0.25, and at X is 0.75 so the entropy is given by $-0.25 \log_2 0.25 - 0.75 \log_2 0.75 = 0.811$. The entropy of all five conditions is tabulated (Table 4.9). No technique provides an entropy of 2, so no technique alone can uniquely characterize all the isomers. However, methods A to C are more effective than D and E. We will return to an analysis of the effectiveness of individual conditions later.

Table 4.9 - Entropy of separation methods in Table 4.8

	Entropy All isomers equally probable	$p_1=0.1$, $p_2..p_4=0.3$
Methods		
A	1.0 (0.5)	0.971 (0.512)
B	1.0 (0.5)	0.971 (0.512)
C	1.0 (0.5)	0.971 (0.512)
D	0.811 (0.406)	0.881 (0.465)
E	0.811 (0.406)	0.469 (0.247)
AB	2.0 (1.0)	1.896 (1.0)
BC	1.0 (0.5)	0.971 (0.512)
AD	1.5 (0.75)	1.571 (0.829)
ADE	2.0 (1.0)	1.896 (1.0)

Note : Figures in brackets indicate proportion of maximum possible entropy for given probabilities.

Since no single technique is able to distinguish all four isomers uniquely, we need to use more than one technique in combination. If we employ both methods A and B, then each isomer is characterized by a unique combination of results. It is easy to calculate the entropy of this combination of results. For example, the entropy for combining methods A and D can be calculated as follows.

(1) Conditions A and D both yielding a compound that elutes in cluster X have a probability of 0.5; $-0.5 \log_2 0.5 = 0.5$.

(2) Condition A yielding a compound that elutes in cluster Y and condition D yielding a compound that elutes in cluster Y have a probability of 0.25; $-0.25 \log_2 0.25 = 0.5$.
(3) Conditions A and D both yielding a compound that elutes in cluster Y have a probability of 0.25; $-0.25 \log_2 0.25 = 0.5$.

Hence combining methods A and D has an entropy of 1.5, which is not quite enough to distinguish all isomers. Methods A, D and E in combination have an entropy of 2 and, therefore, provide enough information to distinguish all four isomers. However, methods A and B together will also distinguish all four isomers uniquely. It is, of course, possible to calculate the entropy of any combination of method(s). The analyst aims to choose a combination of conditions that gives an entropy of 2. Any further conditions result in redundant information. A general strategy in such cases might be to determine the target entropy, and then calculate entropies for any two combinations of methods. If one or more combination(s) are found that meet this target no further combinations of conditions are required. Otherwise combinations of three conditions are tested and so on.

Sometimes, we have more information available about a given system. Let us assume that the probability of isomer 1 occurring is only 0.1, whereas the chances of isomers 2 to 4 occurring are 0.3 for each isomer: the isomers may be extracted from natural sources and we might know that one possible structure is less likely than the other three. In this case the maximum achievable entropy reduces to 1.896, and we see that the relative entropies of each condition and combination of conditions (Table 4.9) change. In particular condition E is now not particularly useful. The factor p_n does not necessarily need to correspond to a probability but could be a weighting function. It might be that we have little interest in isomer 1 or can easily determine isomer 1 by another test, so decide that isomer 1 is less significant. This can readily be incorporated into entropy calculations.

These arguments can be extended to continuous distributions. Consider the problem of separating two isomers again, but this time, instead of discrete peaks, let us model the peaks as two isosceles triangles, with a base width of 2δ (Fig 4.14). If these peaks are completely separated, there will be 1 bit of information. We can use the areas under the triangles to compute information content. We assume the height of each triangle is $1/2\delta$, so the areas are 1, and multiply each area by 0.5 to give probabilities. Obviously we could use any other weighting function as described above, if required. Consider the case when there is some overlap between these triangles. In such a case two compounds are not exactly resolved, so the entropy will be reduced. If the region of overlap is $\rho\delta$ (Fig. 4.14) (i.e. the centres of the peaks are separated by $(2\delta - \rho\delta)$ units), the entropy becomes

$$S \quad = \quad - (0.5+0.5) \, \rho\delta/2\delta \, \log_2 (0.5+0.5) - \; 2 \, (0.5 \, (1-\rho\delta/2\delta) \, \log_2 0.5)$$

$$= \qquad (1 - \rho/2) \, \log_2 2 \quad = \quad (1 - \rho/2) \qquad\qquad (4.35)$$

If there is complete overlap, $\rho=2$, and so the entropy is 0, i.e. there is no longer any information about the separations, whereas if $\rho=0$ (complete separation) the entropy is 1 so there is maximum information.

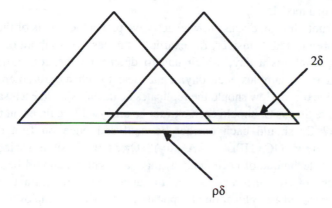

2δ

ρδ

Fig. 4.14 - Two partially overlapping triangular peakshapes as discussed in the text.

Of course this example above is somewhat simple, but there is no reason why more realistic peakshape functions cannot be employed (e.g. Gaussians) or why several peaks and several methods or combinations of methods cannot also be compared in this way.

The information theoretical approach has not been employed greatly for chemometric method selection. One of the weaknesses is the need to model a system effectively, although it is not always necessary to have a target entropy, and it is possible to merely see which techniques increase information content. The approach has certainly been used to compare measurement techniques such as MS and GC, but one of the problems is the need to understand which measurements are correlated. For example, it might be possible to monitor 50 MS peaks. Because most compounds provide a series of peaks, wildly inaccurate estimates of information content would result from treating each MS measurement independently. It is possible to try to simulate and estimate correlations in such cases but this would involve massive computations as there would be a large number of correlations and possible combinations.

4.5 REPLICATING THE SYSTEM

In Chapter 2 we discussed the importance of replicating experiments so as to compute the magnitude of the errors in a system and estimate how well a model is obeyed relative to the experimental error. The reader is referred to Section 2.5 for a discussion of ANOVA for utilizing this information in assessing the confidence in a model.

We did not, though, discuss how exactly we plan which parts of the system need to be replicated. Consider, for example, the case of a laboratory able to handle 100 analyses a day, which has to determine the concentration of compounds in 20 samples each day. That means each measurement can be replicated five times. How should the replicates be taken? Should each sample be extracted five times? Or should each extract be analysed five times on the same instrument? Or should each extract be analysed once on five different instruments (e.g. GC, HPLC, AA etc.)? Or is there such variability in experimental technique of each analyst, that each sample should be analysed once by each of five different technicians using the same extract and same machine? Or should a combination of replication strategies be employed?

We should normally aim to choose the strategy that reduces analytical (experimental) error most. In that way we can then model the results (if necessary) with greater confidence, since the final outcome of our 20 analyses might be a set of numbers that we want to make sense of by using multilinear regression (Section 2.5), PCA (Section 4.2) or a variety of other techniques. Obviously, the better (more certain) our raw data, the more useful the resultant interpretation. The chemometrician must look at the system as a whole and not just isolate data analysis from design.

To study which factors in the measurement process are least subject to error we can again use simple forms of *experimental design* and ANOVA, but in a slightly different way to Chapter 2. In this case, we are not principally interested in building a quantitative model as to how good or bad a technician is in estimating the concentration of a compound, but assessing technician variability against, for example, instrumental variability. This can be estimated through sums of squares of replication errors without the need for detailed model building.

Before analysing these errors, it is necessary, as usual, to design an experiment. There are two principal designs for estimating sources of error in replicate measurement.

The first is called *hierarchical classification*. Consider a situation whereby a sample is given to an analyst, who then extracts the sample and measures a parameter, such as a concentration of a compound. There are at least three sources of *variability* in such a procedure, namely the analyst (different people have different techniques), the extracts, and the variability of the instrumental method. Which source of error is most serious?

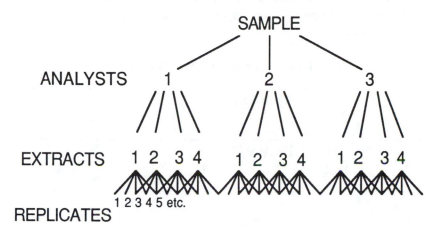

Fig. 4.15 - Hierarchical classification example.

One approach to studying these errors might be as follows.
(1) Give the sample to several analysts.
(2) Each analyst performs several extractions on his sample.
(3) Each analyst measures the concentration of a compound several times on each sample using the same instrument.

There are three *levels* in the process, namely, analyst, extract and instrumental replicate. Each level can be replicated. For an example, consider the case of three ($=I$) analysts who are all given the same sample, who then extract this sample four ($=J$) times, and measure the concentration of a compound five ($=K$) times from each extract. This can be represented as a hierarchical tree (Fig. 4.15). Some simulated data are tabulated (Table 4.10).

Since all analyses are performed on the same sample, we would like to know which of the three factors influences the overall experimental error most. We will call the value of the kth replicate from extract j of analyst i, x_{ijk}. We define the following means.

$$\bar{x}_{ij} = \sum_{k=1}^{k=K} x_{ijk}/K \qquad (4.36)$$

$$\bar{x}_i = \sum_{j=1}^{j=J} \bar{x}_{ij}/J \qquad (4.37)$$

$$\bar{x} \quad = \quad \sum_{i=1}^{i=I} \bar{x}_i/I \quad = \quad (\sum_{i=1}^{i=I} \sum_{j=1}^{j=J} \sum_{k=1}^{k=K} x_{ijk}) \,/\, IJK \qquad (4.38)$$

Eq. (4.35) merely defines the mean of all the analyses together.

We can compute several error sums of squares corresponding to the three means above, namely

Table 4.10 - Hierarchical ANOVA example

ANALYST											
1				2				3			
EXTRACT											
1	2	3	4	1	2	3	4	1	2	3	4
7.987	8.119	9.160	9.300	6.881	6.700	7.302	7.462	8.691	7.374	7.905	7.578
8.877	7.301	9.508	8.793	7.968	6.738	7.570	8.210	8.796	7.679	7.863	7.670
7.841	7.479	9.645	8.221	6.665	7.102	7.231	8.071	8.533	7.806	7.988	8.236
8.606	7.889	8.958	8.743	7.026	7.964	7.028	8.454	8.579	7.765	8.370	7.921
8.127	7.424	9.233	9.119	6.517	8.133	7.614	7.819	9.085	7.607	7.843	7.897

$$S_0 \quad = \quad \sum_{i=1}^{i=I} \sum_{j=1}^{j=J} \sum_{k=1}^{k=K} (x_{ijk} - \bar{x}_{ij})^2 \qquad (4.39)$$

$$S_1 \quad = \quad J \sum_{i=1}^{i=I} \sum_{j=1}^{j=J} (\bar{x}_{ij} - \bar{x}_i)^2 \qquad (4.40)$$

$$S_2 \quad = \quad J K \sum_{i=1}^{i=I} (\bar{x}_i - \bar{x})^2 \qquad (4.41)$$

All these sums of squares are for the full IJK (=3×4×5 = 60) readings in the example, hence the multiplication factors. We need to multiply the sums in Eqs (4.37) and (4.38) by J and JK respectively, as IJK (=60) measurements are available to assess the influence of each of the three factors on the overall error.[1]

_____

[1] S_2 could be calculated by replacing each of the JK (=20) measurements made by each of the I (=3) analysts by the mean reading for each of the analysts, and subtracting these IJK readings (which can now only take one of I values) from the overall mean for all the samples, squaring this difference and summing it. For computational convenience, of course, it is easier to perform I (=3) calculations and multiply the result of each of these calculations by JK (=20).

Table 4.11 - ANOVA of data in Table 4.10

Sum($=S$)	Factors	d.f.($=\nu$)	S	S/ν	$S/\nu=f(\sigma)$
S_0	Sample(k) Extract(j) Analyst(i)	$IJ(K-1)$ $=48$	6.950	0.145	σ_k^2
S_1	Extract(j) Analyst(i)	$I(J-1)$ $=9$	8.183	0.909	$\sigma_k^2+5\sigma_j^2$
S_2	Analyst(i)	$(I-1)$ $=2$	12.068	6.034	$\sigma_k^2+5\sigma_j^2+20\sigma_i^2$

Notes. Sums are as defined in Eqs (4.39) to (4.41).
        The last column can be obtained from Eqs (4.42) to (4.44).

However, the sums of errors in Eqs (4.39) to (4.41) are still due to a combination of factors. To analyse this problem further we need to consider the *degrees of freedom* for each of the three terms above.[1] There are, of course, $IJK-1$ ($=59$) degrees for the overall sum of squares for all the samples corrected for the mean. The degrees of freedom for each of the three terms in Eqs (4.39) to (4.41) is given in Table 4.11. It is easy to verify that they add up to a total of $IJK-1$ degrees of freedom. If we define the root mean square error due to factor m as σ_m, the reader should be able to show that

$$\frac{S_0}{IJ(K-1)} = \sigma_k^2 \qquad (4.42)$$

$$\frac{S_1}{I(J-1)} = K\sigma_j^2 + \sigma_k^2 \qquad (4.43)$$

$$\frac{S_2}{(I-1)} = JK\sigma_i^2 + K\sigma_j^2 + \sigma_k^2 \qquad (4.44)$$

The values of the three root mean sum of square error terms for each factor are computed in Table 4.12. We see that the variability due to using different analysts is more serious than that due to replicate instrumental and extraction error. Thus we have the possibility of a larger undetected error creeping in if the same person analyses a sample several times than if the task is shared out among several people, but each person takes fewer replicates. The significance of the

---

[1]Further discussion of the concept of degrees of freedom is given in Section 2.5.

various sums of squares can be analysed further via the F-test as has already been discussed in Section 2.5, to which the interested reader is referred.

Table 4.12 - Errors calculated from Table 4.11

Factor	σ^2	σ
Analyst (=i)	0.256	0.506
Extract (=j)	0.153	0.391
Samples (=k)	0.145	0.381

A crucial point to note, though, is that the above analysis is strongly dependent on how we choose to study these sources of variability, in other words on the experimental design. A different approach might be as follows.
(1) Extract the sample four times.
(2) Give each of the four extracts to three different analysts (making 12 analysts in total).
(3) Each analyst then performs five replicate analyses.

In such a case the hierarchy differs, with extract at the highest level, analyst at the next level and replicates at the bottom. Yet another situation arises if we design the experiment as follows.
(1) Extract the sample four times.
(2) Give each of the four extracts to three analysts (making three analysts in total).
(3) Each analyst then performs five replicate analysis.

In such a situation, the analysts and extracts are at the same level of the hierarchy. This is a considerably more difficult problem to the one analysed in detail above and much far harder to interpret. As we see below, this latter experiment is a mixture of cross-classification and hierarchical designs. So even simple considerations, such as whether a sample is extracted four times before being given to three analysts or the same sample is given to the three analysts who then extract it can make a major difference to the interpretation of the results and understanding of the sources of error.

A second method of using replicates to assess experimental errors is called *cross-classification*. In the terminology introduced above, we say that all the factors are at the same level. Consider the following experiment.
(1) A sample is extracted four (=J) times.

(2) Each extract is given to three (=I) analysts.
(3) Each analyst performs one measurement on five different (=K) gas chromatographs.

There are now three factors, namely, extract, analyst and gas chromatograph, but now all three factors are of equal importance. Of course, if we knew above, that the first replicate was performed on one machine, and the second on another machine and so on, we could have used this foreknowledge to improve the analysis of errors, but we were not told this and assumed that there was no significance in the order in which replicates were taken. With greater knowledge of a system our data analysis methods must change as is usual in chemometrics.

Some simulated replicates data obtained using a cross-classified design are tabulated in Table 4.13. Although this table may appear superficially similar to Table 4.10, in fact it is very different. First, the rows actually are significant. In Table 4.10 the order in which the replicates were tabulated was not significant. Second, we do not need to put analysts on the top level; indeed, we could completely reorganize the table. An alternative way to represent the experiment may be as cells in a three dimensional box (Fig. 4.16) with each axis representing one factor. A third very important difference is that there are now *four* and not three sources of error. There are the three systematic sources, namely analyst (=i), extract (=j) and instrument (=k) and finally a random error. We could, of course, replicate each single instrumental analysis yet again. If we did this three times, we would need to take 180 and not 60 measurements, which involves a large amount of time and effort. Furthermore, the resultant ANOVA analysis would be quite complex. There would be three equal sources of error at the top of a hierarchy and a replicate error at the bottom of the hierarchy, so such the experiment is a two level hierarchical experiment, with the top level itself a three factor cross-classification experiment. The analysis of such experiments is outside the scope of this text but potentially the subject of many original research papers in chemometrics. In fact in the data of Table 4.13, we have already taken 60 measurements so the pure (replicate) error is buried within the data.

It is possible, for factor i, to compute

$$S_i \quad = \quad JK \sum_{i=1}^{i=I} (\bar{x}_i - \bar{x})^2 \qquad (4.45)$$

using notation of Eqs (4.36) to (4.38). These equations will be symmetrical for all 3 factors, as they are all of equal levels. It can then be shown that

$$S_i/(I-1) \quad = \quad \sigma_0^2 + JK\sigma_i^2 \qquad (4.46)$$

since the sum in Eq. (4.45) has ($I-1$) degrees of freedom. We denote the error for factor i as σ_i and the analytical (replicate) error as σ_0. The replicate error can be estimated as follows,

$$\sigma_0^2 = (\ [\sum_{i=1}^{i=I} \sum_{j=1}^{j=J} \sum_{k=1}^{k=K} (x_{ijk}-\bar{x})^2] - (S_i+S_j+S_k)\ ) / (IJK-1-(I+J+K-3)) \qquad (4.47)$$

Table 4.13 - Cross-classification ANOVA example

	ANALYST											
	1				2				3			
	EXTRACT											
	1	2	3	4	1	2	3	4	1	2	3	4
INSTR.												
1	7.840	8.293	8.427	7.835	8.846	9.147	8.269	8.837	7.048	9.797	8.125	9.651
2	5.922	7.691	7.081	7.390	8.464	8.495	8.679	8.563	8.402	9.347	8.737	9.049
3	6.926	7.413	7.107	8.941	6.856	8.197	7.935	8.447	7.092	9.221	9.239	10.065
4	6.053	6.704	7.162	8.184	8.083	8.704	8.970	8.556	8.319	7.886	8.330	8.070
5	6.119	7.234	7.912	8.390	8.117	8.972	8.166	8.662	7.981	9.316	8.068	8.726

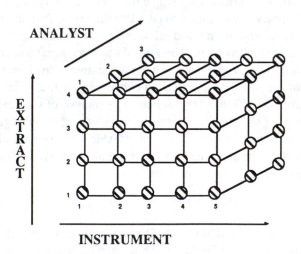

Fig. 4.16 - Cross-classification example.

This equation may look fairly complex (there is insufficient room in this text to prove everything in detail) but is actually relatively easy to understand. The residual sum of squares due to unsystematic error is the difference of the overall variance (or squared deviation from the mean) of the system minus the three systematic sum of square terms divided by the residual degrees of freedom (i.e. $59 - 4 - 3 - 2 = 50$ in our example). The overall variance for our example is 47.814. The three values of S_m are calculated in Table 4.14, giving, a value for $\sigma_0^2 = 0.352$, and three other systematic error terms (due to analyst, extraction and instrumental variability) as tabulated. As can be seen, the error due to instrumental variability is less than that due to extraction which is less than that due to analyst variability. This type of information can aid the laboratory manager plan his sampling strategy. The various sums of squares can, of course, also be compared using F-tests as described in Section 2.5.

Replicates analyses can be used in further situations to examine and minimize sources of error. For example it is possible for the response to be *multivariate* rather than univariate, e.g. the intensity of 20 chromatographic peaks. A method called multivariate analysis of variance (MANOVA) can be employed to analyse these errors. As yet relatively few chemometricians use such approaches. This is probably because classical chemometrics was centred around a few fairly specialized methods such as partial least squares and SIMCA (described in Chapters 6 and 7) and chemometricians have only recently turned their attention to more widely used statistical methods.

Table 4.14 - Calculation of errors for data in Table 4.13

Factor (=m)	S_m	ν	S_m/ν	$S_m=f(\sigma)$	σ_m^2	σ_m
Analyst (i)	16.578	2 (I−1)	8.289	$\sigma_0^2+20\sigma_i^2$	0.396	0.630
Extract (j)	11.424	3 (J−1)	3.808	$\sigma_0^2+15\sigma_j^2$	0.230	0.480
Samples (k)	2.190	4 (K−1)	0.5475	$\sigma_0^2+12\sigma_k^2$	0.016	0.128

Therefore, by planning the way replicates are recorded, the chemometrician can maximize the quality of his or her data for the same amount of effort. In

Chapter 2, we discussed the importance of having replicate information so as to compare errors in the model to those of the experimental process. In this section we have shown how these errors can be reduced by carefully considering which elements of an experimental procedure are most error prone. Clearly the lower the analytical (replicate) error the more confidence there is in a model, so it is useful to be able to decide in advance of extensive experimentation which parts of the system are least reproducible.

There is a large statistical literature on comparing errors in interacting processes. In this section we restrict the discussion entirely to comparison of variance (i.e. errors about the mean) but there are other tests which involve comparing means rather than variances. For example, one person or apparatus might consistently over- or underestimate a parameter: this sort of error is often quite easy to sort out. Also there are sophisticated tests for directly comparing three rather than two errors (by an F-test). In most practical situations, though, the laboratory based chemist wants to ask fairly qualitative questions, such as whether an extraction procedure or using different measurement techniques introduces the most serious errors into the system. Obviously it is up to the chemometrician how seriously he analyses each step in the experimental procedure, and this, yet again, depends on the time and resources available to him. In this text we recommend that the chemometrician tries to sort out which errors in an experimental system are largest but the *significance* of these errors must be balanced out against non-statistical factors. Consider a routine procedure involving three replicate analyses. The only decision is whether to extract a sample three times, or to use three different technicians or to measure the same extract using the same technician three times and so on. Statistical significance of the relative errors is of little interest. The main question we need to ask is which part of the system is most error prone? In other cases the time spent taking replicates might need to be balanced against the time spent processing the data; the number of samples that can be analysed in a given time may be constrained by the number of instruments and technicians available; funds might become available which could be used to employ extra technicians or to purchase new equipment or to employ a programmer to analyse data further. These decisions involve looking at the entire experimental system as a whole and it is impossible to provide general guidelines.

4.6 IMPROVING INSTRUMENTAL PERFORMANCE

Once an analytical technique has been chosen, using methods outlined in Sections 4.2 to 4.4, and the need for and nature of replicate analyses is ascertained, it is necessary to *optimize* the instrumental method. Some of the theory of experimental design and optimization has been outlined in Chapter 2.

In this section we briefly review how such approaches can be employed to improve instrumental performance.

Perhaps one of the most well established examples is *chromatographic optimization*. A chromatographic response function (CRF) which is normally a univariate parameter such as the average signal height of the peaks in a chromatogram is chosen: a sequential or simultaneous experiment is designed to improve this parameter. For example, if the CRF is the average peak width in a chromatogram, conditions are chosen that minimize this, so improving resolution. There is a vast literature on choice of CRF which we do not have space to discuss in this text.

Most literature concerns the optimization of HPLC separations. A simple problem involves a three component mixture of solvents. Seven experiments, arranged as in Table 2.9, can be performed; each corner of the triangle corresponds to a pure solvent. At each combination of solvents the CRF is measured, and the resultant data are used to fit a polynomial which, in turn, is used to determine the optimum of the response surface. However this type of design ignores some basic information available to the chromatographer. In addition to solvent composition the *eluotropic strength* of a solvent also determines its ability to separate compounds. There are a number of means of measuring eluotropic strength but the most cited work is due to Snyder (see bibliography at end of this chapter). He measures a value S for the eluotropic strength of common solvents. The value for water is defined as 0; we will use values for methanol of 2.6, acetonitrile of 3.1 and THF (tetrahydrofuran) of 4.4.[1] The overall eluotropic strength of a mixture of I solvents is given by

$$\gamma = \sum_{i=1}^{i=I} S_i \phi_i \qquad (4.48)$$

where ϕ_i is the proportion of solvent i. Eq. (4.48) assumes that the influence of a given solvent on a separation is linearly proportional to its presence in the solvent mixture. This is true for most solvents, but *not necessarily* for added reagents (e.g. ion pairing reagents) and in some cases the specific effect of adding certain reagents on the eluotropic strength of a solvent mixture must be studied and non-linear terms will appear in Eq. (4.48). One difficulty with a straight mixture design of the form of Table 2.9 is that the eluotropic strength at each experimental point differs. For example if the three solvents are methanol, acetonitrile and THF, the eluotropic strength will vary between 2.6 and 4.4 in the design. It is, of course, possible to model the influence of eluotropic strength on the CRF as well as solvent composition (so γ becomes an extra parameter), but this reduces the number of degrees of freedom. Some chromatographers prefer to perform all experiments at the same eluotropic strength. For example, if

---

[1]There are several differences according to authors; these depend on the experiment used to measure these parameters.

we want to set $\gamma=1$, then the maximum proportion of methanol becomes $1/2.6 = 0.38$ rather than 1. Since the proportion of solvents no longer add up to 1, it is necessary to introduce a fourth solvent, water, which has $S=0$. The proportion of water is then given by

$$\phi_{water} = 1 - \phi_{THF} - \phi_{acetonitrile} - \phi_{methanol} \qquad (4.49)$$

A simplex centroid mixture design is given in Table 4.15.

Table 4.15 - Simplex centroid design for constant eluotropic strength

Expt	%Methanol	%Acetonitrile	%THF	%Water
1	38.4	0	0	61.6
2	0	32.3	0	67.7
3	0	0	22.7	77.3
4	19.2	16.1	0	64.7
5	19.2	0	11.4	69.2
6	0	16.1	11.4	72.5
7	12.8	10.8	7.6	68.9

Note : These experiments correspond to those of Table 2.9 with methanol as factor A, acetonitrile as factor B and THF as factor C, and a constant eluotropic strength of 1 as described in the text.

However, many chromatographers routinely change the gradient during an experiment. This approach was particularly important in classical chromatography where two solvent delivery systems were usual. Three or four solvent delivery systems are now normally available in most modern HPLC equipment. Hence much attention has recently been placed on three or four solvent mixture designs. But for difficult separations gradient chromatography is still useful. Indeed it is important not to get too enthusiastic about the use of standard chemometric techniques if these do not allow us to incorporate information that we are virtually sure can help. Many chromatographers intuitively know a gradient will improve separation for a given problem so it is not very sensible to use a mixture design that does not take gradients into account. There has been relatively little literature on gradient mixture designs, but one approach is to use a mixture "prism". The conditions at one end of the prism represent an eluotropic strength of γ_1 and the other end represent conditions of γ_2. It is easy to compute these conditions for any values of γ as in Table 4.15. A prism is illustrated in Fig. 4.17. The experiment is normally performed over a given time (e.g. 60 minutes) and seven experiments involve changing the solvent composition linearly over time from the point in one triangle to the corresponding point in the other triangle. For example, if factor A is pure methanol and we want $\gamma_1=0.5$ and $\gamma_2=2.0$, then the proportion of methanol at the start of a run is 20.8% (= $100 \times 0.5 / 2.4$) and at the end of the run is 83.3% (= $100 \times 2.0 / 2.4$) for Expt 1 (corresponding to Table 2.9). More elaborate gradient mixture designs can be proposed. One of the weaknesses of such designs is that the gradient is fixed, in other words γ_1 and γ_2 are constant for each experiment. There remains further work in the area of variable gradients, but these involve rather complex designs, not as yet reported in the chemometric literature.

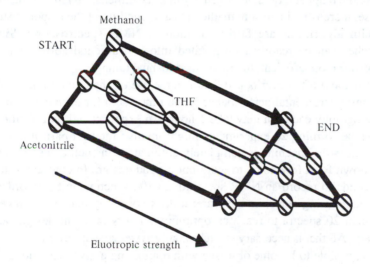

Fig. 4.17 - Gradient mixture design.

Many people use simplex methods (see Section 2.3) for the optimization of instrumental techniques such as HPLC, GC and AA. Several weaknesses of the simplex approach have already been discussed. A major practical problem involves defining the factor to be optimized – the progress of the simplex depends on what response function is chosen. For simultaneous designs it is possible to choose any response function after the experiments have been performed and model the resultant data as discussed in Section 2.5. Simplex methods are also rather harder to automate than factorial approaches. It is easy to hand a technician a set of conditions which he can then carry out without referring back after each experiment; similarly a computer can readily be programmed to carry out a mixture design. For simplex optimization, someone, or some computer, must work out, after each experiment, what next experiment to perform. In the case of chromatography the choice is often non-trivial as it depends, in turn, on choice of response functions.

Simplex approaches are, though, effective under certain circumstances and a classical example is the tuning of scientific instruments, such as NMR spectrometers: one peak (for example, the signal due to deuterium in the solvent in ^{1}H NMR) is chosen, and the magnetic field is adjusted to minimize the width of this peak. Although it is possible to perform such experiments manually, it typically takes a few seconds or less to record the width of such a peak, so several hundred spectra can be recorded in a few minutes. The response is a simple univariate one, each experiment is fast and there are several factors (for example the magnetic field homogeneity in three dimensions) that influence the response; there is no interest in mathematical modelling of the response surface. Autoshim algorithms are fairly common in NMR spectroscopy. Simplex approaches can be readily incorporated into instrumental software and are probably far more efficient, timewise, than manual tuning.

A final and often overlooked factor in the use of chemometric methods for improving instrumental performance is the time taken for each experiment. In chromatography each run may take 1 hour. An experimental optimization that involves performing 20 experiments might involve the best part of a working week, unless the machine is completely automated. In such a case it is worth sitting down for a few hours to work out a good design. In contrast, it takes a few seconds to record an NMR spectrum so 100 spectra can be recorded in a few minutes using autoshim software: it does not really matter if it takes 100 rather than 20 spectra to reach an optimum, so very efficient designs are not necessary. All that is necessary is a reasonably systematic approach.

It is possible to become obsessed with optimizing a given technique. There are no precise guidelines as to how much effort is needed to improve a given instrumental method. Probably it is best to determine limits on the time spent in this pursuit. If these limits look like being exceeded careful evaluation of the pros and cons of investing further resources needs to be made. If a project is sufficiently important it may be worth investing several months' effort; otherwise it is probably best to use a partially optimized system and to return to the problem of instrumental optimization if the results (e.g. quality of spectra, separation or quantification of compounds) are not good enough. In many cases data processing (for example methods for resolving overlapping peaks) is complementary to experimental design. A balance must be made according to the nature of a project and resources, in terms of manpower, finance and instrumentation.

4.7 ACQUIRING INFORMATION MORE EFFICIENTLY USING FOURIER AND HADAMARD SPECTROSCOPY

Most of this text is about how best to deal with *signals* in the presence of *noise* or experimental error. By processing data effectively the trends buried within the noise can be distinguished from the noise. But chemometrics is also about using time and resources with maximal efficiency, and the level of noise in a dataset can be reduced by changing the method used to acquire the data. In Sections 4.3 and 4.4 we discussed how to choose the most useful analytical technique and in Section 4.5 we discussed the importance of understanding the main sources of noise (or error) and planning the experimental (or replicate) strategy to make best use of such information. The next section outlined how chemometrics can improve instrumental methods for acquiring data. However, we did not take into account the time required to record each spectrum or chromatogram. Consider a situation where procrustes analysis (Section 4.3) suggests that one HPLC run is as informative as one NMR and one mass spectrum together for a particular problem. But if one HPLC run takes 1 hour, whereas an NMR spectrum can be set up, tuned and recorded in 10 minutes and a mass spectrum in 15 minutes, which approach is most useful now? Once conditions are established (e.g. the NMR sample is prepared and inserted in the NMR machine) replicates may take only a few seconds or minutes, so, within 1 hour it may be possible to record 20 replicate NMR spectra and one non-replicated mass spectrum. Is this new strategy better than just recording a single HPLC run? If two skilled technicians and expensive instrumentation are required to record the NMR and MS information, whereas one technician and cheaper instrumentation are needed to record the HPLC information it is still possible that the HPLC approach is better. But if the *signal to noise ratio* of the HPLC is very poor, and several replicates are required in order to produce acceptable quantitation for a particular problem, and each replicate takes 1 hour to record, compared to a few seconds for an NMR replicate, NMR and MS combined may now become the methods of choice.

An added factor, not considered above, in our choice of instrumental method, is how *efficiently* each method acquires data. We want to improve the amount of information acquired per unit time. Although it is, indeed, possible to use measures of information content based on entropy (discussed in Section 4.4), below we adopt a somewhat simpler approach. We will restrict the discussion in this section to spectroscopy. In Section 4.2 we introduced the concept of signal to noise ratios. We noted that if we average the information from two precisely correlated peaks the signal to noise ratio improves by a factor of $\sqrt{2}$. The reasoning of Eqs (4.2) to (4.6) can be extended and it can be shown that for most symmetrical noise distributions the signal to noise ratio increases by $\sqrt{N}$ if N correlated peaks are recorded. Instead of measuring N peaks in a single spectrum, why not record the same spectrum N times? Each time a spectrum is recorded the signals exactly overlap so the signal strength is linearly related to the number of scans, whereas the noise does not exactly coincide and is related to the square root of the number of scans. If we add the N spectra together the signal to noise ratio increases as is illustrated in Fig. 4.18. This is the principle of *signal averaging*. The two variables illustrated Fig. 4.6 could represent the intensity of the same peak recorded twice rather than two peaks arising from the same source measured once. There is no mathematical difference.

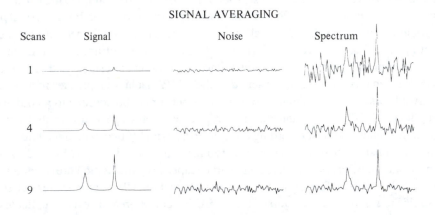

SIGNAL AVERAGING

Fig. 4.18 - Principles of signal averaging.

The approach of *signal averaging* then, provides us with better information, as measured by signal to noise ratios, and is one way of improving the quality of instrumental data. Signal averaging can, nevertheless, be a time consuming process. For example, if it takes 5 minutes to scan one spectrum, it takes 500 minutes, or about 8 hours, to scan 100 spectra and so improve the signal to noise ratio 10-fold. An alternative approach might be to find a method whereby 100 peaks arising from the same source are monitored simultaneously: in this way each time a spectrum is recorded there will be 100 correlated peaks corresponding to one factor. There are a few spectroscopic techniques such as pyrolysis MS where a compound of reasonable complexity may give rise to 10 or 20 peaks, but very few techniques where a single compound gives rise to 100 peaks.

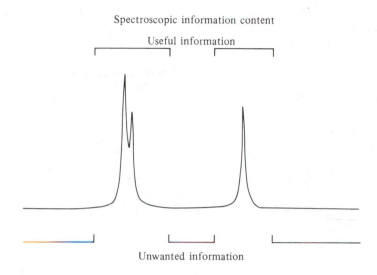

Fig. 4.19 - Information content of an idealised spectrum. Areas between peaks are of no interest to the spectroscopist.

So different approaches are required. We should consider the efficiency of the spectroscopic process. Fig. 4.19 illustrates a symbolic spectrum: the region between the peaks is of no direct interest to the spectroscopist. The peaks might only occupy 20% of the spectrum, so 80% of the time the spectroscopist is recording noise between the peaks. This is surely a most inefficient use of instrument time. If the spectroscopist knew, in advance, exactly where the peaks were, he or she could, of course, modify the experiment so as to spend the majority of time recording information in these regions of the spectrum and scan rapidly over the regions between the peaks. As is usual in chemometrics, the more that is known about the system, the more the possibilities of optimizing the system, and improving the signal to noise ratio in a given time. But if the spectroscopist knew everything about the system in advance, there would be no point recording the spectrum: normally we only have a small amount of prior knowledge of the system. Simple knowledge might extend to typical peakshapes or rough concentrations. Generally the spectroscopist has fairly precise knowledge of the range of spectroscopic frequencies that are of interest. For example in ^{1}H NMR spectroscopy, resonances generally occur within a range of

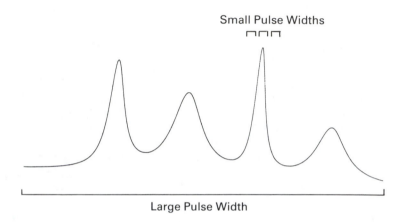

Small Pulse Widths

Large Pulse Width

Fig. 4.20 - Use of a large pulse width in Fourier spectroscopy.

10 parts per million (ppm) with respect to the magnetic field. Occasionally there may be peaks of interest outside this range, but chemical knowledge is generally used to decide whether there are peaks outside the normal frequency range. Therefore, it is only necessary to observe a very small fraction of possible frequencies. Within this experimental region peaks might occur almost anywhere, so rather than scan the whole spectrum from beginning to end, it is more efficient, normally, to observe the entire spectrum at once. This can be visualized as increasing the slit width from, for example, one thousandth of the spectrum, to the entire spectrum (Fig. 4.20). The conventional, scanning, spectrometer, might sample the spectrum at 1000 different frequencies. The spectroscopist, however, usually wants to know information such as, *3 peaks of intensity 5, 8 and 12 are centred at frequencies 185, 543 and 610 Hz* rather than the intensities at observations 1 to 1000, so requires rather specialized information from the spectrometer. Instead of obtaining such information in 1000 observations, each single observation is a function of all the information in the spectrum. Clearly, though, he or she must record several such observations to disentangle all the spectroscopic information. But his new method of

recording *all the data all the time* is far more efficient than the scanning spectrometer. This approach is the basis of Fourier spectroscopy.

Table 4.15 - Correspondence between frequency and time domains in Fourier spectroscopy

Time domain	Frequency domain
Decay mechanism (e.g. exponential)	Lineshape (e.g. Lorentzian)
Decay rate	Line width
Initial intensity	Integral
Frequency	Position

Normally since each datapoint recorded by a Fourier spectrometer is a function of the entire spectrum, it is not easy to interpret directly the raw experimental data. The directly interpretable spectrum consists of a sum of peaks, normally characterized by a position, width, area and shape. For Gaussian peakshapes (see Fig. 3.1) the position is equivalent to μ and width to σ and the area $(= A)$ is related to the height and width.[1] Thus the spectrum could be regarded as a sum of functions of the form

$$x_i = \sum_{n=1}^{n=N} f(A,\mu,\sigma,n) + \varepsilon_i \qquad (4.50)$$

where x_i is the intensity at datapoint i, ε_i is a noise function and there are N peaks in the spectrum. Normally this type of display is called the *frequency domain*. However the raw data from the Fourier spectrometer are not in such an easily interpretable form, and a *Fourier transform* is used to convert this to a frequency domain spectrum. Formally these raw data are called a *time series* and are the directly observed output from a Fourier spectrometer. The data normally consist of a sum of sine waves characterized by a frequency, initial intensity, decay constant and decay mechanism. Each component corresponds to a frequency domain peak. The correspondence between parameters in both domains in given in Table 4.15. The process of transformation between time and frequency domains is illustrated in Fig. 4.21. The mathematical basis of Fourier transformation is discussed in greater detail in Section 5.2.

In order to acquire Fourier data, it is necessary to use special spectroscopic techniques. The two major forms of Fourier spectrometer are *pulsed* and

---

[1]Lineshapes are discussed in greater detail in Section 5.6.

interferometric methods. In the former approach all the spectral frequencies are excited in one short pulse, and the decay of the excited spectrum is followed with time: NMR is a good example. In the latter, optical methods are used: IR is one example, where a Michelson interferometer is employed. It is important to recognize that, although Fourier transform instrumentation could be built for most forms of spectroscopy, there has to be commitment on the part of manufacturers to build such spectrometers. In spectroscopies such as NMR and IR where such instruments have been commercially available for almost two decades, routine access is normally possible in most laboratories. In contrast, although uv / visible spectrometry also has potential as a form of Fourier spectroscopy relatively few moves have been made towards commercialization: this is partly due to market demand.

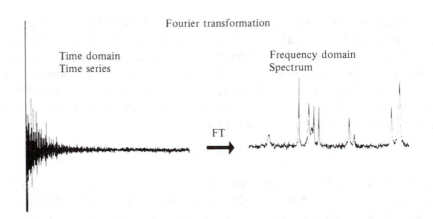

Fig. 4.21 - Principles of Fourier transformation.

Another form of spectroscopy is called *Hadamard spectroscopy*. If we want to observe the intensity of radiation at I wavelengths, it is better, as described above, to observe several wavelengths at a time. One simple approach is to plan M experiments at each of which certain wavelengths are omitted, so that the net intensity for experiment m is given by

$$\xi_m = \sum_{i=1}^{i=I} s_i x_i + \varepsilon_m \tag{4.51}$$

where s_i takes a value of either 1 or 0, x_i is the true intensity at wavelength i and ε_m is a noise function. The advantage of this approach is that the signal to noise ratio is increased. If the M experiments are planned properly it is possible to work out the values of x_i at each of the wavelengths. A simple experiment where M and I both equal 2 involves

(1) Observing the summed intensity at both wavelengths ($s_1 = s_2 = 1$)

(2) Observing the intensity at one wavelength only ($s_1 = 1$, $s_2 = 0$).

These experiments clearly need to be planned properly so that we can deduce all the information about the original spectrum. If experiment 2 involved setting $s_1 = s_2 = 0$, we would be unable to recover the original information.

The experiments can be expressed in matrix form, whereby

$$\mathbf{Y} \quad = \quad \mathbf{S} \cdot \mathbf{X} + \quad \mathbf{E} \qquad (4.52)$$

In this case, $\mathbf{Y}$ is a vector of length M, $\mathbf{S}$ is called an $\mathbf{S}$ matrix which is a matrix of order $M \times I$, $\mathbf{X}$ is a vector of length I containing the true intensities at each of the wavelengths and $\mathbf{E}$ contains residual errors. If the experiments are planned correctly is is possible to obtain all relevant information in I experiments. Thus $\mathbf{S}$ becomes a square matrix. We will restrict the discussion below to square matrices.

The $\mathbf{S}$ matrix is exactly analogous to experimental design matrices discussion in Chapter 2. The design in Hadamard spectroscopy consists of performing M (=I in the case discussed below) experiments each of which involves recording one response which is the sum of intensities at P wavelengths where $P < I$. These experiments are then used to provide the experimenter with a noise free estimate of x_i. There is a great deal of theory behind the construction of $\mathbf{S}$ matrices, which we will not discuss in detail here, but they relate to Hadamard matrices ($\mathbf{H}$ matrices). These are matrices consisting of elements whose value is either -1 or $+1$, but the product of any two rows is always 0, so that

$$\sum_{i=1}^{i=I} c_{ni} c_{mi} \quad = \quad 0 \qquad (4.53)$$

where c_{mk} has the value $+1$ or -1 and is the element of the kth column and mth row. A *normalized* $\mathbf{H}$ matrix is one in which the first element of each column and row takes the value of $+1$. It can be shown that $\mathbf{H}$ matrices have an order of 2, 4, or a multiple of 4. $\mathbf{S}$ matrices are derived from $\mathbf{H}$ matrices by

(1) removing the first column and row (so take on the order 1, 3, or $4n-1$) and

(2) changing the -1s to $+1$s and $+1$s to 0s.

A matrix of order 7 represents seven experiments performed at 7 different wavelengths. Each experiment measures the total spectroscopic intensity at a sum of those wavelengths for which $s_i = 1$ in the $\mathbf{H}$ matrix. This matrix can be regarded as a *mask*: those wavelengths with value 0 are masked, usually by a physical block.

Table 4.17 - Construction of a 7×7 cyclical **S** matrix

Normalized **H** matrix

1	1	1	1	1	1	1	1
1	-1	-1	-1	1	-1	1	1
1	1	-1	-1	-1	1	-1	1
1	1	1	-1	-1	-1	1	-1
1	-1	1	1	-1	-1	-1	1
1	1	-1	1	1	-1	-1	-1
1	-1	1	-1	1	1	-1	-1
1	-1	-1	1	-1	1	1	-1

Resultant cyclical **S** matrix

1	1	1	0	1	0	0
0	1	1	1	0	1	0
0	0	1	1	1	0	1
1	0	0	1	1	1	0
0	1	0	0	1	1	1
1	0	1	0	0	1	1
1	1	0	1	0	0	1

One problem with this approach is that it involves building quite complex and elaborate instrumentation if a large number of wavelengths are involved. If we want to observe 999 wavelengths, we need 999 masks. So normally spectroscopists employ *cyclical S matrices*. In these each row is shifted by one column relative to the row above so that $c_{(n+1),(i+1)} = c_{ni}$. In such a case only one mask is needed, and that mask is merely shifted along the spectrum. Normally for I wavelengths the mask is $2I-1$ elements in length, with elements $I+1$ to $2I-1$ identical to elements 1 to $I-1$. The mask for the first experiment is given by elements 1 to I, the second by elements 2 to $I+1$ and so on.

The construction of a 7×7 cyclical **S** matrix is illustrated in Table 4.17. Each row consists of an experiment, so that the third row would be interpreted as "observe wavelengths 3, 4, 5 and 7", for example. The mask is of the form 1101001110100 and by shifting this mask by one unit, each experiment, given by each row of the matrix in Table 4.17, is carried out.

It is, in fact, very easy to compute the best spectrum from the results of the Hadamard experiment. The values x_i for the N wavelengths may be expressed in matrix form and are given by

$$\mathbf{X} = \mathbf{S}^{-1} \mathbf{Y} \qquad\qquad (4.54)$$

where $\mathbf{Y}$ is a vector of the intensities for the N experiments.

Hadamard spectroscopy has been less popular than Fourier spectroscopy because of the problems of constructing rapidly changing masks, but has recently been successfully applied in several areas, most notably Raman spectroscopy.

REFERENCES

General

D.L.Massart, A.Dijkstra and L.Kaufman, *Evaluation and Optimization of Laboratory Methods and Analytical Procedures*, Elsevier, Amsterdam, 1978

G.Kateman and F.W.Pijpers, *Quality Control in Analytical Chemistry*, Wiley, New York, 1981

R.Caulcutt and R.Boddy, *Statistics for Analytical Chemists*, Chapman and Hall, London, 1983

Section 4.2

K.V.Mardia, J.T.Kent and J.Bibby, *Multivariate Analysis*, Academic Press, London, 1979

C.Chatfield and A.J.Collins, *An Introduction to Multivariate Analysis*, Chapman and Hall, London, 1986

S.Wold, K.Esbensen and P.Geladi, *Chemometrics Int. Lab. Systems*, **2**, 37 (1987)

I.T.Joliffe, *Principal Components Analysis*, Springer, Berlin, 1986

E.Malinowski and D.Howery, *Factor Analysis in Chemistry*, Wiley, New York, 1980

O.M.Kvalheim, *Anal. Chim. Acta*, **177**, 71 (1985)

Section 4.3

J.C.Gower in *Mathematics in the Archaeological and Historical Sciences*, F.R.Hodson, D.G.Kendall and P.Tautu (Editors), University Press, Edinburgh, pp. 138-147 (1971)

W.J.Krzanowski, *Appl. Statist.*, **36**, 22 (1987)

Section 4.4

D.L.Massart, *J. Chromatogr.*, **79**, 157 (1973)

Section 4.5

See general references to this chapter and Chapter 2.

Section 4.6

J.C.Berridge, *Techniques for Automated Optimization of HPLC Separations*, Wiley, New York, 1985

P. Schoenmakers, *Optimization of Chromatographic Selectivity*, Elsevier, Amsterdam, 1986

J.C.Berridge, *Chemometrics Int. Lab. Systems*, **5**, 195 (1989)

L.R.Snyder and J.Kirkland, *Introduction to Modern Liquid Chromatography*, Wiley, New York, 1979

Section 4.7
R.G.Brereton, *Chemometrics Int. Lab. Systems*, **1**, 17 (1986)
A.G.Marshall (Editor), *Fourier, Hadamard and Hilbert Transforms in Chemistry*, Plenum, New York, 1982
R.R.Ernst, *Adv. Magn. Reson.*, **24**, 271 (1966)

5

Univariate signal processing

5.1 UNIVARIATE VERSUS MULTIVARIATE APPROACHES TO SIGNAL PROCESSING

Signal processing lies at the centre of chemometrics. Nearly all laboratory data are acquired by computerized instruments. From these machines parameters such as the number and position of peaks in a spectrum, their relative integrals, and sometimes their shapes, are computed, and are the key to the interpretation of large masses of data. Hands-on spectroscopists and chromatographers have realized the importance of enhancing the quality of instrumental signals for many years, and software is routinely available on most computerized instruments. Yet many chemometricians resist using methods for signal processing. Many chemometricians became interested in the subject *via* multivariate statistics and have taken the trouble to learn the basis of matrix algebra, ANOVA, PCA and so on, and are, therefore, often unfamiliar with or reluctant to use simple digital filters or Fourier transforms as part of their data analytical strategy. This psychological resistance is no different to the resistance many experimenters have to using statistical experimental designs. Yet methods for univariate signal processing can help improve the quality of chemical data, and lead to better and more efficient interpretation of experiments: nearly all modern laboratory data are acquired by computerized instruments. The ultimate goal of chemometrics is to reach scientific truths as efficiently and confidently as possible and the chemometrician must look at all steps in the system and not be prejudiced in favour of any particular method.

Univariate approaches discussed in this chapter are complementary to multivariate methods discussed in Chapter 6. Generally if a set of instrumental signals is available, such as in the case of diode array HPLC, multivariate approaches like factor analysis or partial least squares are the methods of choice. If individual spectra or chromatograms are available, univariate approaches such as those described in this chapter are preferred. It is, of course, possible to mix both types of approach. For example, signals might be smoothed to reduce noise prior to factor analysis. It is very important to recognize that these classes of method are complementary.

Univariate approaches to signal processing represent probably one of the first applications of on-line laboratory computing. There is a massive literature in this area, probably considerably larger than the rest of the literature on chemometrics. This area is also of considerable interest to engineers and instrument builders and much detailed work was reported before cheap on-line computing became readily available. Simple filters and transformations can readily be performed by electric circuits, and the resultant filtered signal may be displayed on an oscilloscope. The disadvantage with electronic filtering is that it is not possible to return later and recalculate the spectrum: there is only one chance and so the original emphasis was on cautious, on-line, methods. Although there is a battery of statistical methods for time series analysis and filtering, these often require elaborate computational calculations and knowledge of signal shape or noise distributions, not generally available on-line. Therefore, most of the early signal processing software failed to exploit knowledge about signals and noise but safely carried out rapid, simple, signal enhancement. With the possibility of extensive on-line computing power, with second processors, with large, rapidly accessed, disc storage, and with facile data transfer between machines it is now possible to apply more sophisticated chemometric methods for signal processing in a routine manner. There are still problems convincing the hands-on spectroscopist or technician that thinking about the nature of signals and noise will make them more productive. Spectroscopists are especially used to believing in their spectrum. Traditionally, spectroscopists calculate parameters such as the energy of bonding of two atoms from exact spectral frequencies. The problems of signal processing are rarely considered in traditional physical chemistry. The early users of on-line computing in chemistry were mainly physical chemists using techniques such as NMR and IR spectroscopy. They were principally interested in deriving exact physical models: as discussed in Chapter 1 exact models have an important place in chemistry but cannot really be considered a part of chemometrics. In the 1970s vast libraries of physical data, such as second order spin simulations in NMR spectroscopy, were built up. Much of this work is now virtually redundant with the advent of better instrumentation (for example, in NMR it is usual to use high field strengths, decoupling, two dimensional spectroscopy and so on rather than elaborate modelling of second order couplings, and the elaborate quantum mechanical software routines of the early 1970s are rarely used as a tool for the practicing experimental spectroscopist). Instrument builders took many of their ideas from sound engineering principles and there was very little interest in developing elaborate tests for noise distributions and so on.

The development of commercial signal processing software is a very time intensive and so expensive task. Unfortunately market forces are more important than chemometric theory. Most routine laboratory instrumentation is used on a daily basis by technicians rather than statisticians. Criteria such as user-friendliness and rapid, automated, turnover of samples are generally of higher priority than the ability to perform elaborate statistical calculations. Technicians rarely enjoy being told that there is low confidence in their results, even though this type of information is often highly useful. In a busy laboratory rapid results are often more important than chemometrics. The laboratory manager rarely has the time to analyse the significance of a spectrum or to invest time and resources into a study of factors that influence the nature of the noise and peakshape distributions. Hence instrument manufacturers, although often well aware of the problems of signal processing, have concentrated on the production of user-friendly software. Features such as menus, editors, interactive fast graphics, spreadsheets and so on are all part of modern instrumental software. A technician will normally choose the manufacturer that produces the best graphics rather than the manufacturer that produces the most comprehensive statistical diagnostics. The difficulty with market driven software design is that crucial information about the quality of the data and confidence in conclusions is often missing or hard to surmise. The instrument manufacturer will find it more cost effective to hire a programmer to produce a mouse driven menu system than an ANOVA package.

There needs, therefore, to be considerably greater emphasis placed on the chemometric approach to signal processing.

5.2 AN INTRODUCTION TO FOURIER TRANSFORMS

We have already discussed cyclical time series (Section 3.4) and Fourier transform instrumentation (Section 4.7). In this section we briefly outline the basis behind Fourier transform methods. Many of the definitions in this section will be used later. The reader already familiar with Fourier transforms may omit this section.

As discussed previously (Section 4.7), directly interpretable information is normally said to be contained in the *frequency domain*.[1] Corresponding to the

---

[1] There is some confusion between the spectroscopic and chromatographic literature, but we define the frequency domain as the domain of direct chemical interest, even if it is recorded as a time profile (chromatography) rather than a frequency profile.

frequency domain is a *time domain*, generally consisting of a sum of decaying sine waves, as we show below. These two domains are interrelated by the process of Fourier transformation (Fig. 4.21). In previous sections we introduced Fourier transform instrumentation as a means of acquiring data more efficiently. In the same way that multivariate measurements are more informative than univariate measurements, especially if several peaks are measured simultaneously, so Fourier (and Hadamard) transform spectrometers are built so that several intensities in a spectrum are acquired simultaneously. However, Fourier transformation is a general method of data-processing and can be applied to any dataset, in whatever way obtained. In this chapter we will discuss Fourier transforms in the context of signal processing. There is no need to acquire data in the time domain in order to process data by Fourier methods. Every frequency domain has a corresponding time domain. So Fourier transform techniques can be applied to any sequential dataset. Under such circumstances these methods are used for computational convenience and not to increase efficiency of data acquisition.

Several signal processing methods discussed below are best applied in the time domain, so we need to be conversant with Fourier methods and to be able to change between time and frequency domain whenever we wish.

There are two basic forms of Fourier transformation. These are defined as *cosine transforms*

$$RL(\zeta) = \int_{-\infty}^{\infty} f(t)\cos\zeta t\, dt \tag{5.1}$$

and *sine transforms*

$$IM(\zeta) = \int_{-\infty}^{\infty} f(t)\sin\zeta t\, dt \tag{5.2}$$

where $f(t)$ is the value of the time domain at time t and $RL(\zeta)$ and $IM(\zeta)$ the value of the frequency domain at frequency ζ in cycles per unit time.[1] Some authors prefer to use radians per unit time, in which case $\omega = 2\pi\zeta$. For simplicity we will adopt cycles per unit time (1 Hz = 1 cycle s^{-1}), but the units must be carefully noted when reading the literature. Various texts use radians s^{-1}, in which case the definitions in this section need to be modified by multiplying by various factors, normally $(1/2\pi)$ for either the forward or inverse transform or $\sqrt{(1/2\pi)}$ for both transforms. Note that there are two results of transforming the

---

[1] Some authors use v rather than ζ for frequency. We avoid this in this text because v is used as a noise function elsewhere.

frequency domain. Normally these are combined together into what we call a *forward transform*:[1]

$$F(\zeta) \quad = \quad RL(\zeta) - iIM(\zeta) \quad = \quad \int_{-\infty}^{\infty} f(t)e^{-i\zeta t}dt \tag{5.3}$$

We can similarly define *inverse transforms* which are given by

$$rl(t) \quad = \quad \int_{-\infty}^{\infty} F(\zeta)\cos\zeta t d\zeta \tag{5.4}$$

and

$$im(t) \quad = \quad \int_{-\infty}^{\infty} F(\zeta)\sin\zeta t d\zeta \tag{5.5}$$

or

$$f(t) \quad = \quad rl(t) + iim(t) \quad = \quad \int_{-\infty}^{\infty} F(\zeta)e^{+i\zeta t}dt \tag{5.6}$$

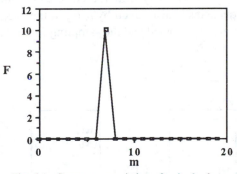

Fig. 5.1 - Spectrum consisting of a single sharp spike.

Chemists, however, rarely deal with continuous datasets. Normally they sample a spectrum or chromatogram at regular time or frequency intervals, and over a finite interval in time. To cope with this data, we use *discrete Fourier transforms* (DFTs). The discrete form of Eq. (5.4) is, for example,

$$rl(n\delta t) \quad = \quad \sum_{m=0}^{m=M-1} F(m\delta\zeta)\cos(n\delta t m\delta\zeta) \tag{5.7}$$

---

[1] $\cos\zeta t - i\sin\zeta t = e^{-i\zeta t}$ and $\cos\zeta t + i\sin\zeta t = e^{+i\zeta t}$ where $i = \sqrt{-1}$.

where δt is the sampling interval in time and $\delta\zeta$ in frequency. In this case M points are sampled in the frequency domain. The corresponding inverse transform is normally multiplied by a factor of $(1/M)$ to keep the magnitude of the numbers the same when a forward transform is followed by an inverse transform, or *vice versa*. Sometimes both transforms are multiplied by $\sqrt{(1/M)}$ for purposes of symmetry.

A simple numerical example will be used to illustrate the principles of Fourier transformation. Consider a spectrum consisting of a single spike (Fig. 5.1) centred at $m = 7$, sampled over 20 points (from $m = 0$ to 19), of height 10. The values of $F(m\zeta) = 0$ if $m \neq 7$. Then,

$$rl(n\delta t) = \qquad 10\cos(7\,\delta t\,n\delta\zeta) \tag{5.8}$$

or a simple cosine wave. Normally, if M points are sampled, the inverse transform preserves the overall number of datapoints so that there are $M/2$ real and $M/2$ imaginary points, for computational convenience. Algorithms can be developed so that the first half of the transform contains the values for $rl(n\delta t)$ and the second half for $im(n\delta t)$ as illustrated in Fig. 5.2.[1] In Section 3.4 we discussed the problem of *Nyquist frequency* when sampling cyclical time series. The reader is referred to Fig. 3.9 where we demonstrate that the range of distinguishable frequencies is given by $1/2\delta t$ for a discretely sampled time series. But this range of frequencies is also given by $M\delta\zeta$ in the case discussed here, since M frequency points are sampled and the sampling frequency is $\delta\zeta$. Hence,

$$M\delta\zeta \quad = \quad 1/(2\delta t) \tag{5.9}$$

Substituting in Eq. (2.8) we have

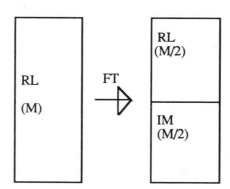

Fig. 5.2 - Typical organization of data in a Fourier transform algorithm.

[1]There is a large literature on algorithms for fast Fourier transformations. We do not have space in this text to discuss these methods in detail but refer the reader to references in the bibliography if interested.

$$rl\ (n\delta t)\ =\ 10 \cos (7\ n\ /\ M) \qquad (5.10)$$

Since use units of cycles s^{-1} (so $\cos(1) = 1$), the real transform peak in (5.10) will be a cosine wave with 3.5 cycles in the transform as illustrated in Fig. 5.3. It should be evident, then, that a spike of intensity I_0 centred at $\mu\delta\zeta$ transforms to a real cosine wave of the form

$$rl\ (n\delta t)\ =\ I_0 \cos (n\ \mu/M) \qquad (5.11)$$

In other words a peak centred 38 datapoints away from the origin transforms into a cosine wave with 19 cycles (or maxima) in the transform. The corresponding imaginary transform merely is the corresponding sine wave.[1]

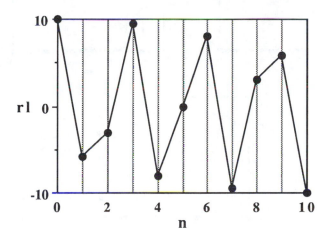

Fig. 5.3 - Real transform of data in Fig. 5.1.

Peaks are not, however, infinitely sharp: in chromatography different molecules have a range of adsorption properties and in spectroscopy molecules relax slowly. Much of signal processing is concerned with the problems of broad peaks, which we will discuss in detail in later sections. A better model for a peak may be a triangle several datapoints wide. A series of three triangles of increasing width is illustrated in Fig. 5.4. The transform now becomes

$$rl(n\delta t) = I_0[\cos(n\mu/M) + \sum_{k=1}^{k=T-1}([T-k]/T)\{\cos(n(\mu-k)/M) + \cos(n(\mu+k)/M)\}] \qquad (5.12)$$

---

[1] There are a number of other definitions differing in units, and datalengths and the reader must carefully note this before reading texts and papers on Fourier transforms.

where T is the half width of the triangle in datapoints (i.e. 1 for the top triangle, 2 for the next and 3 for the third). The data, together with the real transforms are given in Table 5.1 and the transforms are illustrated in Fig. 5.5.

There are several important features of these transforms, which can be verified by the more mathematically minded readers.

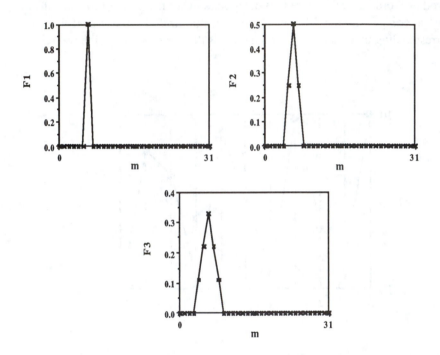

Fig. 5.4 - Raw data for the three triangular functions of Table 5.1.

(1) The curve is always a cosine wave of frequency μ.
(2) The initial datapoint of the cosine wave is proportional the integral of the original peak, which is 1 for the three triangles above. This is easy to verify, since the integral is merely $\delta\zeta(1+2((T-1)/T)+2((T-2)/T+...)$ which is the same as the result of putting $n=0$ in Eq. (5.12) apart from the factor $\delta\zeta$.
(3) The cosine wave decays if the peak is broader than one datapoint in width. In fact the *broader the peak, the faster the corresponding transform decays*.[1]

---

[1] There is a slight increase in the relative intensity for the third peak towards the end of the transform. This is due to other small effects which will not be discussed here, but which the interested reader can verify.

Table 5.1 - Discrete Fourier transforms of triangular functions

Observation n (or m)	Raw data $F_1(m\delta\zeta)$	$F_2(m\delta\zeta)$	$F_3(m\delta\zeta)$	Fourier transform $rl_1(n\delta t)$	$rl_2(n\delta t)$	$rl_3(n\delta t)$
0	0	0	0	1.000	1.000	1.000
1	0	0	0	0.195	0.193	0.190
2	0	0	0	−0.924	−0.889	−0.832
3	0	0	0	−0.556	−0.509	−0.438
4	0	0	0.111	0.707	0.604	0.458
5	0	0.25	0.222	0.831	0.647	0.411
6	1.0	0.5	0.333	−0.383	−0.265	−0.133
7	0	0.25	0.222	−0.981	−0.586	−0.211
8	0	0	0.111	0.000	0.000	0.000
9	0	0	0	0.981	0.395	0.041
10	0	0	0	0.383	0.118	−0.002
11	0	0	0	−0.831	−0.185	−0.001
12	0	0	0	−0.707	−0.104	−0.013
13	0	0	0	0.556	0.047	0.027
14	0	0	0	0.924	0.035	0.074
15	0	0	0	−0.195	−0.002	−0.020
16	0	0	0			
17	0	0	0			
18	0	0	0			
19	0	0	0			
20	0	0	0			
23	0	0	0			
24	0	0	0			
25	0	0	0			
26	0	0	0			
27	0	0	0			
28	0	0	0			
29	0	0	0			
30	0	0	0			
31	0	0	0			

Real transform according to Eq. (5.12). Some authors display the imaginary transform in points 16 to 31.

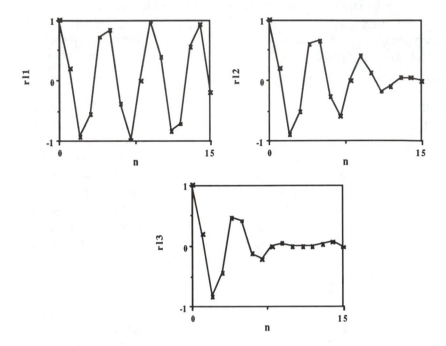

Fig. 5.5 - Fourier transform of the three triangular functions illustrated in
Fig. 5.4.

These results can be generalized, and the reader is referred to more specialized texts for detailed mathematical discussion. Table 4.15 compares frequency and time domain parameters. We will discuss detailed signal shapes in Section 5.6 below.

It is important to remember that it is not always necessary to transform from a *real* frequency domain to a *complex* (real and imaginary) time domain, but sometimes from a real time domain to a complex frequency domain. The former case is common in chromatography where data are usually obtained in the frequency domain (using definitions above); the latter is usual in Fourier transform spectroscopy (see Section 4.7) such as NMR.

Consider transforming M purely real time domain points using a forward transform of the form Eq. (5.3) to $M/2$ real and $M/2$ imaginary frequency domain points. In such a case there will be both real and imaginary transforms. The real transform of a decaying cosine wave is often called the *absorption spectrum* whereas the imaginary transform is the *dispersion spectrum*. These are illustrated in Fig. 5.6. Normally we are only interested in the absorption spectrum.

Real and imaginary pairs

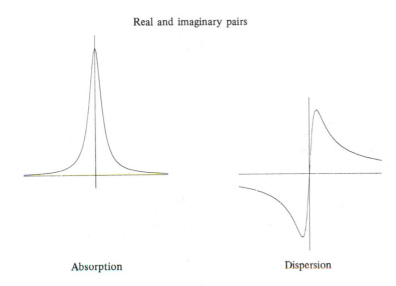

Absorption   Dispersion

Fig. 5.6. Absorption and dispersion spectra.

Sometimes the first datapoint is not exactly at a maximum. This occurs if there is a short delay between the signal to acquire data and a computerized instrument actually acquiring such data as illustrated in Fig. 5.7. This delay means that the real time series is not exactly a cosine wave, but often of the form

$$f(t) \quad = \quad g(t)\cos((\omega+\phi)t) \tag{5.13}$$

where $g(t)$ is a non-cyclical decay function (as discussed in later sections of the text) and ϕ is called the *phase angle*. As illustrated in Fig. 5.7, this phase angle is dependent on the delay time and the frequency of the sine wave. In the resultant transform the absorption and dispersion are mixed up, so that the real spectrum does not exactly correspond to the absorption spectrum. The appearance of out of phase peaks is illustrated in Fig. 5.8. It is possible to recover the absorption and dispersion spectra by phasing so that

$$ABS(\zeta) \quad = \quad \cos(\phi)\, rl(\zeta) + \sin(\phi)\, im(\zeta) \tag{5.14}$$

and

$$DIS(\zeta) \quad = \quad \sin(\phi)\, rl(\zeta) + \cos(\phi)\, im(\zeta) \tag{5.15}$$

where $ABS(\zeta)$ and $DIS(\zeta)$ are the absorption and dispersion spectra respectively. There are many methods for determining the value of ϕ.

There are many other causes of phase errors beside the first order effects illustrated in Eq. (5.13) and Fig. 5.7, and these can be modelled by including

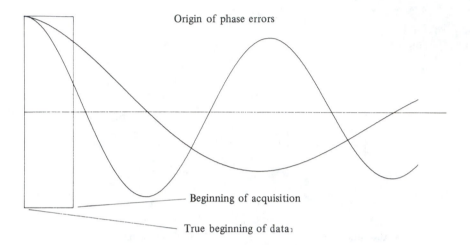

Origin of phase errors

Beginning of acquisition

True beginning of data

Fig. 5.7 - First order phase errors caused by delay times.

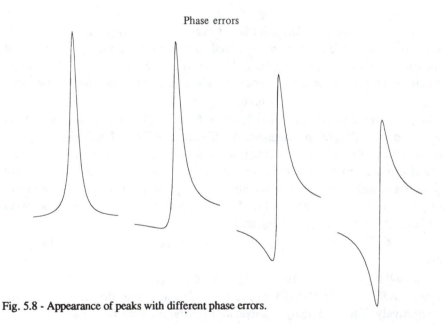

Phase errors

Fig. 5.8 - Appearance of peaks with different phase errors.

more angular terms in Eq. (5.13). Sometimes, such as in three dimensional imaging, where the image (e.g. NMR or X-ray) is obtained by passing radiation over a complex object, the causes of phase imperfections can be hard to understand as the phase errors often depend on angle and on the constitution of the material. Under such circumstances, Eq. (5.13) may be replaced by a polynomial dependent on orientation, and the best fit polynomial is the one that recovers the most positive intensities. Finding this can be quite hard but modern approaches to image processing, such as maximum entropy, discussed below, can help.

Finally, it is not always necessary to display the real or imaginary spectrum, but the power spectrum defined by

$$P(\zeta) \quad = \quad rl^2(\zeta) + im^2(\zeta) \qquad\qquad (5.16)$$

or the absolute value spectrum (also called the magnitude spectrum) defined by

$$M(\zeta) \quad = \quad \sqrt{P(\zeta)} \qquad\qquad (5.17)$$

These spectra do not require the data to be phased. The weakness is that lineshapes are somewhat distorted and the spectra are non-quantitative. The latter problem is an important one to the chemometrician. The area under two overlapping peaks does not equal the sum of areas for each individual component for the simple reason that $(a+b)^2 \neq a^2 + b^2$. It would destroy the objective of good experimental design and sophisticated pattern recognition if non-quantitative data were used in later stages of the analysis. This illustrates the importance of considering all aspects of the experimental process, including the way in which Fourier transform data is displayed, when analysing a system.

The discussion on Fourier transforms could be expanded considerably, but the most important definitions used below are given in this section. It is necessary to be extremely careful when reading a paper on Fourier transforms. There are a huge number of possible conventions and definitions. Important points to note are as follows.

(1) Value (if any) of constant in front of the Fourier integral or summation.
(2) Whether the frequency or time domain is purely real or consists of real / imaginary pairs.
(3) Whether a real dataset of M datapoints is transformed to a real dataset of M real datapoints or to a real dataset of $M/2$ datapoints and imaginary dataset of $M/2$ datapoints.
(4) Whether frequency is in cycles s^{-1} or radians s^{-1}.
(5) Whether the definitions of inverse and forward transforms are as above or have been reversed.
(6) Whether the frequency or time domain is directly interpretable.

In the discussion below we do not use a constant in front of integrals for continuous transforms, but use a constant of $(1/M)$ in front of summations for discrete inverse transforms only; normally we will be converting a real time domain to real and imaginary frequency domains; we will convert a real domain of M datapoints to a real and imaginary domains of $M/2$ datapoints each; the frequency will be in cycles s^{-1}; we will adhere to the definitions of inverse and forward transform given above; the frequency domain will be the directly interpretable domain even if it is recorded in time.

5.3 MOVING AVERAGES AND LINEAR DECONVOLUTION

We return to the problem of signal processing. One of the most important uses of chemometrics is to enhance apparent signal to noise ratios. Consider one of the commonest functions of a computerized instrument which is to provide a printout which is a list of peaks in a chromatogram or spectrum, with their positions and intensities or integrals. Most people who have developed software to perform this function (often called "peakpick" algorithms) recognize that this task is far from trivial, because, as always, *noise* is imposed upon *signals*. In a noise free system, one method for deciding what peaks are present might be to scan the data, and if, $x_{i-1} < x_i > x_{i+1}$ where x_j is the intensity at position j, then i is considered to be the position of a peak. This is illustrated in Fig. 5.9. A more realistic criterion might be to involve 5 points in the calculation so that $x_{i-2} < x_{i-1} < x_i > x_{i+1} > x_{i+2}$. Neither of these criteria would work well in the case of partially overlapping peaks, but we will discuss this again below. One of the problems with these methods, however, is that they do not take into account noise distributions. Fig. 5.10 illustrates the same series of peaks as in Fig. 5.9

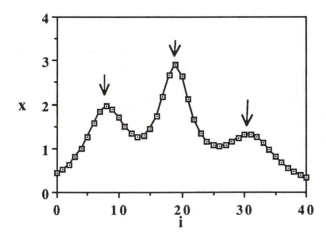

Fig. 5.9 - Simple 3 point peakpick approach identifies three peaks in a noise-free spectrum.

but in the presence of noise, and it is easy to see that the three point peakpick algorithm now also picks noise as well as true peaks. One approach to improving the performance of the peakpick algorithm would be to acquire less noisy data, by optimizing instrumental conditions, using Fourier or Hadamard methods, averaging replicate runs or whatever is appropriate in the particular case (see Chapter 4). Sometimes this will work, but in other cases this approach is time consuming or not possible (e.g. because of unstable compounds, limited instrumental availability) so the chemometrician must choose between data acquisition (experimental design and optimization) and data processing (data analysis). In this chapter we are principally concerned with data processing. An alternative approach to using Fourier and related spectroscopic techniques to enhance signal to noise ratios is using computational methods to reduce the noise in the spectrum or chromatogram. Obviously both instrumental methods (building better instruments) and computational approaches are complementary.

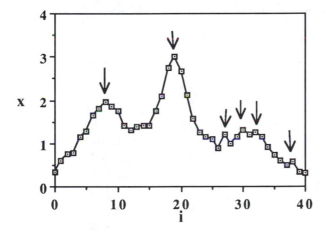

Fig. 5.10 - Data of Fig. 5.9 in the presence of a small amount of noise. The 3 point peakpick approach now tries to identify six peaks.

A chemometrician puts partial knowledge of a system to use. For example he or she often knows that real spectroscopic peaks can be described by functions such as Gaussian and Cauchy distributions (see Section 5.6 for further details). These functions are continuous and systematic. Noise, on the other hand, generally is uncorrelated, that is the value of the noise at sampling point i is not dependent on the value at sampling point $i+1$: this type of noise is referred to as *white noise*. Occasionally there is some correlation between successive values of noise (e.g. ARMA processes as discussed in Section 3.2: this is referred to as *coloured noise*) but the noise function is still not a smooth predictable distribution. One of the simplest approaches to enhancing signal to noise ratio is to use a *moving average* filter. A linear moving average is given by

$$\hat{x}_i = (1/\sum_{j=-N}^{j=N}\lambda_{|j|})\sum_{j=-N}^{j=N}\lambda_{|j|}x_{i+j} \qquad (5.18)$$

This is called a 2N+1 point moving average. The values of $\lambda_{|j|}$ are weights. An unweighted 5 point moving average filter will be given by

$$\hat{x}_i = (1/5)(x_{i-2}+x_{i-1}+x_i+x_{i+1}+x_{i+2}) \qquad (5.19)$$

for example where $\lambda_i = 1$ for all five datapoints. It is possible to devise triangular or other weights. Table 5.2 illustrates the calculation of a 5 point linear unweighted moving average. It should be noted that N end points are deleted from the resultant averaged data.

Table 5.2 - Five point unweighted moving average filter

i	x_i	$\hat{x}_i$
1	0.509	-
2	0.351	-
3	0.912	0.778
4	0.809	1.068
5	1.311	1.463
6	1.955	1.843
7	2.330	2.164
8	2.808	2.369
9	2.414	2.348
10	2.338	2.130
11	1.849	1.791
12	1.241	1.440
13	1.112	1.039
14	0.662	0.733
15	0.281	0.579
16	0.370	0.454
17	0.472	0.310
18	0.485	0.318
19	-0.059	-
20	0.322	-

There is an alternative way of looking at filters where $\lambda_{|j|} = 1$, and that is as a local best fit straight line of the form

$$\hat{x}_{j+i} \quad = \quad b_0 + b_1(j+i) \tag{5.20}$$

We could reformulate the filter as follows.

(1) Centre the filter on point i.
(2) Compute the best fit straight line including points from x_{i-N} to x_{i+N}.
(3) Replace x_i by the best fit value $\hat{x}_i$.

This approach involves computing the least squares best fit local polynomial of order 1. But linear approximations are somewhat unrealistic in many situations, and it is sometimes preferable to increase the order of the model so that, for example,

$$\hat{x}_{i+j} \quad = \quad b_0 + b_1(i+j) + b_2(i+j)^2 + \; ..... \tag{5.21}$$

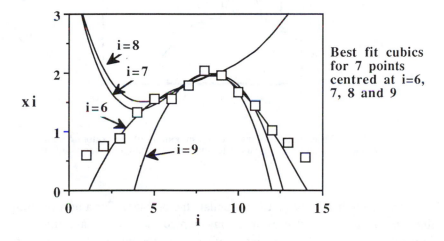

Fig. 5.11 - Best fit 7-point polynomials of order 3 for a typical peak.

Best fit 7-point local polynomials of order 3 for a simple peakshape are illustrated in Fig. 5.11, together with the new values of $\hat{x}_i$. The new values are no longer the unweighted mean, so more complex computational procedures are required to estimate these. Fitting a 990 5-point least squares best fit polynomials to a chromatogram consisting of 1000 experimental points would, however, be an extremely slow process. Early chromatographers and spectroscopists expected on-line answers, but had limited on-line computing power. To have to wait an hour for a chromatogram to be smoothed using a local polynomial would be considered an unacceptable use of time.

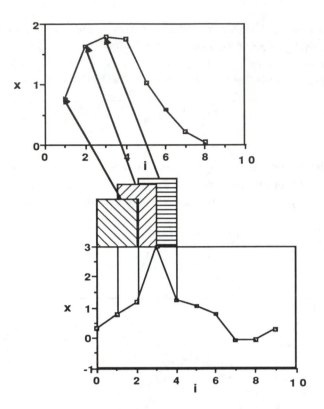

Fig. 5.12 - Three point moving average smoothing of the bottom data to give the top picture is equivalent to applying a rectangular convolution function.

Fortunately it is possible to reformulate the problem. For a best fit straight line, we found, above, that using an unweighted moving average filter of the form of Eq. (5.18) with $\lambda_{|j|} = 1$ was equivalent to computing least squares best fit straight lines: this is because we are only interested in the points in the centre of the polynomial, and not the outside. It is equivalently possible to show that higher order polynomials can be computed using a weighted moving average of the form of Eq. (5.18). This is the principle behind *Savitsky-Golay* filters, originally proposed in 1964. The coefficients for $2N+1 = 5$-, 7- and 9-point smoothing filters (where appropriate) of orders 2, 3, 4 and 5 are tabulated in Table 5.3. Other filters can be obtained from the literature. The advantage of this approach is that a problem of polynomial curve fitting is changed into a problem of weighted moving average filters, which is computationally easier to formulate. These filters are particularly common in chromatography, where exact models of peakshapes (e.g. whether they are Gaussian or Lorentzian) are often not available, but linear moving averages are inadequate.

Table 5.3 - Savitsky-Golay coefficients

	Quadratic / cubic			Quartic / quintic	
No. points	5	7	9	7	9
j					
−4	−	−	−21	−	15
−3	−	−2	14	5	−55
−2	−3	3	39	−30	30
−1	12	6	54	75	135
0	17	7	59	131	179
+1	12	6	54	75	135
+2	−3	3	39	−30	30
+3	−	−2	14	5	−55
+4	−	−	−21	−	15
Normalization	35	21	231	231	429

Note : To compute the coefficients divide by the normalization constant.

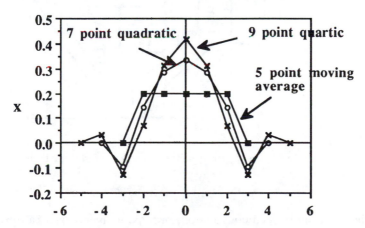

Fig. 5.13 - Illustration of 3 convolution functions.

An equivalent way of looking at moving average functions is as a form of *convolution*. A moving average may be represented as a rectangular convolution function which is gradually moved along the spectrum (Fig. 5.12). The weights given in Eq. (5.18) are equal to the intensity of the convolution function. A 5-point convolution function has weights 0.2 for $j=-2,-1,0,1,2$ and 0 for other values of j. A general convolution operation can be written as

$$\hat{x}_i = \sum_{j=-N}^{j=N} x_{i+j} \times f_j \qquad (5.22)$$

where f_j is a convolution function. This function can be given by a table of weights as the Savitsky-Golay coefficients are or as a mathematical function such as an exponential. The convolution functions for a 5-point moving average, a 7-point local best fit quadratic polynomial function and a 9-point local best fit quartic polynomial function are illustrated in Fig. 5.13. These functions are also sometimes called *windows*.

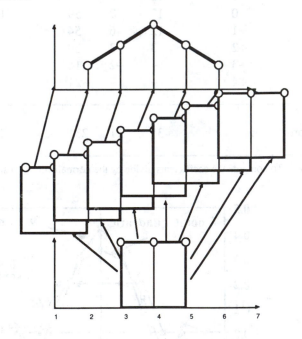

Fig. 5.14 - Convolution of a rectangle (below) with itself to give a triangle (see Table 5.4).

Above we introduced convolution, or moving averages, as a means for reducing noise in chromatograms or spectra. But they can equally well be used to enhance resolution. In Fig. 5.14 we illustrate the convolution of a rectangular peak with another rectangle, of the form $\lambda_{|1|} = \lambda_{|0|} = 0.333$. We see that the shape of the resultant peak changes to a triangle. The calculation is illustrated in Table 5.4. It should be noted that the area under the peak remains constant. Convolution can also be regarded as a method for changing peakshapes. What this means is that convolution can be used to resolve blurred peaks. In spectroscopy or chromatography peaks are blurred as discussed above. The blurring process is illustrated in Fig. 5.15: two sharp peaks are "broadened" by normal physical processes. If we know something about this process of broadening we can surely put this knowledge to use and help recover the original (sharp) spectrum (or chromatogram – for simplicity we will restrict the discussion below to spectra). The chemometrician is principally interested in the sharp spectrum, which may be a list of peaks and their intensities; the sharp spectrum has both higher signal to noise ratio and is better resolved.

Table 5.4 - Convolution of a rectangle to a triangle

i	x_i	$\hat{x}_i$
1	0	0.00
2	0	1.33
3	4	2.67
4	4	4.00
5	4	2.67
6	0	1.33
7	0	0.00

However we could plan how best to *deconvolute* a spectrum if we knew what the peakshape was in advance. A rectangular peakshape, as discussed above, is rather unusual, but more likely peakshapes such as Gaussians can be used. The more knowledge there is of peakshapes the more powerful the deconvolution function. There is an enormous literature on deconvolution but the optimal approach depends heavily on foreknowledge, i.e. how much is known, in advance, about the data. Obviously if everything is known in advance, there is little point using chemometric methods for enhancing the data, so chemometricians have to put partial knowledge to best use. In some techniques, such as solution NMR, the spectroscopist often has detailed and physically reproducible information about peakshapes. In other cases, such as chromatography, this type of information is hard to obtain reproducibly and often depends on a variety of instrumental factors.

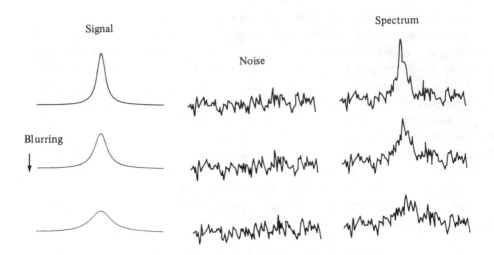

Fig. 5.15 - Illustration of peak blurring or broadening.

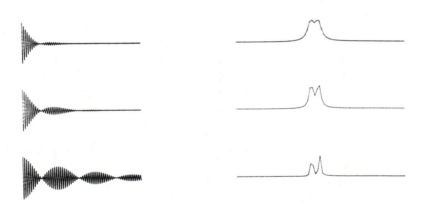

Fig. 5.16 - Result of multiplying a time domain signal (left) by exponentials
of increasing value of λ and the corresponding transforms.

Convolution can also be expressed as a continuous process of the form

$$f'(\zeta) = \int_{-\infty}^{\infty} f(\zeta).g(\zeta-\zeta')d\zeta' \qquad\qquad (5.23)$$

where $f(\zeta)$ is the original (blurred) function, $g(\zeta)$ is a convolution function and $f'(\zeta)$ is the new deconvoluted function. Alternatively we can use the *convolution operator* ($\otimes$) defined by

$$f'(\zeta) \quad = \quad f(\zeta) \otimes g(\zeta) \tag{5.24}$$

Earlier we introduced convolution as a time and labour saving method. The Savitsky-Golay method was much faster than fitting 990 local polynomials, for example. Yet more complicated convolution operations can be slow, particularly if $g(\zeta)$ is defined over the whole spectrum, but we can speed this computation up further. If $F(t)$ is the Fourier transform of $f(\zeta)$, $G(t)$ of $g(\zeta)$, and $F'(t)$ of $f'(\zeta)$

$$F'(t) \quad = \quad F(t) \cdot G(t) \tag{5.25}$$

in other words, convolution in one domain is replaced by multiplication in the other domain. This is the *convolution theorem*. Multiplying by a simple function such as an exponential or a rectangular notch is quicker than performing 1000 integrals, for example. Fig. 5.16 demonstrates the result of multiplying time domain data (as defined in Section 5.2) by a function of the form $e^{+\lambda x}$ where λ is a positive number. The resultant peaks in the frequency domain are now sharper and better resolved. The choice of deconvolution methods will be discussed in greater detail in Section 5.6.

It is, of course, important to choose whether to work in the time or frequency domain. It is always possible to Fourier transform any sequential dataset, and the domain should be the one in which the convolution can be performed fastest. Generally if there is only limited knowledge of peakshapes and noise distribution it is easiest to work in the frequency domain. This is because the convolution functions are normally simple linear moving averages, such as rectangular windows or Savitsky-Golay filters: these often only involve computing a few coefficients. If the chemometrician has rather more elaborate knowledge of peakshapes (e.g. that they are Gaussian) it is often best to work in the time domain. This is because peakshapes can be defined over infinite limits, and it is important to put this knowledge to use, but this involves more elaborate deconvolution functions in the frequency domain with correspondingly slower calculations and so it is much quicker to work in the time domain where the deconvolution operation now merely corresponds to multiplication.

There are a huge number of different types of linear filters, in both the time domain and frequency domain. The method of weighted moving averages can be built upon further.

One approach is an iterative approach called the *van Cittert method*. This is involves computing successively better approximations of f(ζ) by

$$\hat{f}^{n}(\zeta) \quad = \quad \hat{f}^{n-1}(\zeta) + f(\zeta) - \hat{f}^{n-1}(\zeta) \otimes \hat{f}^{n-1}(\zeta) \tag{5.26}$$

where f^{m} refers to the *m*th estimate of the spectrum. Another filter, most usually employed in the time domain is called a *Wiener filter*. In this case, if $N(t)$ is the noise, and $F(t)$ the data in the absence of the noise,

$$G(t) = \frac{|F(t)|^2}{|F(t)|^2 + |N(t)|^2} \qquad\qquad (5.27)$$

This approach deblurs peaks, i.e. produces maximally sharp peaks, providing the original noise and noise-free data are known. Of course, a guess may be made and so partial knowledge of the system put to use.

The chemometrician should carefully pursue the literature of a particular analytical technique. For example, chromatographers often prefer least squares smoothing methods such as the Savitsky-Golay approach. NMR spectroscopists use a variety of filters such a convolution difference, double exponentials, sine bells and so on. Most NMR literature is based on the classical work of Ernst. Uv / visible spectroscopists prefer approaches such as van Cittert's method. Statisticians have developed windows such as the Hanning window (which is a three point moving average $\lambda_{|1|} = 0.25$, $\lambda_{|0|} = 0.5$). The different terminology in each area and the choice of window / deconvolution function reflects partly the different nature of the problems encountered, and the relevant knowledge of noise, lineshapes etc., but also the historical division in the literature. A person writing a paper on NMR filters is likely to get his paper reviewed by another NMR spectroscopist and so has to place his ideas within the framework of NMR spectroscopy: editors are likely to insist that the author changes his paper to fit in with currently accepted terminology employed by NMR spectroscopists if he submits to a specialized spectroscopic journal. There is a strong resistance in signal processing to unifying the various methods in different branches of the subject. Probably, with a more widespread acceptance of chemometric thinking this will change in the future.

5.4 MAXIMUM ENTROPY NON-LINEAR DECONVOLUTION

One of the difficulties of the linear methods for deconvolution described in Section 5.3 is that they have a limited ability to take into account extra knowledge about a system. For example, it might be well known that all peaks in a spectrum or chromatogram have positive intensity. This is a simple form of knowledge. However, noise does not have to be positive in sign. We could, of course, perform linear deconvolution and then decide to replace all values of $\hat{x}_i$ less than 0 by 0 as we know that there are no physical solutions with negative intensity. This, though, is rather a messy solution and the solution to the problem can no longer be expressed by a simple linear combination of coefficients. There is also often other knowledge about the system, such as the number of peaks in a system. It might be known that a chromatogram of a mixture consists of exactly 10 peaks: we want a method for signal processing that identifies no more and no less than 10 peaks and gives us their intensities and positions. The observed chromatogram will be blurred and noisy compared to the estimated chromatogram. In some forms of spectroscopy even more detailed knowledge might be available. For example, in NMR every coupling in a spectrum is related to another coupling elsewhere. Thus the pattern of peaks follows certain well established rules. The spectroscopist or chromatographer intuitively employs this type of information.

Non-linear deconvolution involves methods for enhancing the resolution and / or signal to noise ratio that cannot be expressed analytically as linear sums of the form of Eq. (5.18). Probably the most developed method is based on *maximum entropy*. We introduced the concept of entropy in Section 4.4, and will build on this below.

A spectrum may be considered as a probability density function. Normally we measure some spectroscopic information and try to build up a fuller picture of the underlying system using chemometric methods.

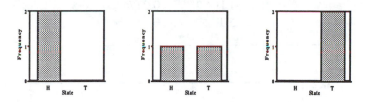

Fig. 5.17 - Histograms of three possible outcomes of the toss of two unbiassed coins.

A simple non-chemical example is the toss of two unbiassed coins. There are three possible outcomes, viz.
(1) 2 Heads (H)
(2) 1 H and 1 Tail (T)
(3) 2 T
as illustrated in Fig. 5.17. Which of the three experimental outcomes is most likely? Intuitively we expect 1 H and 1 T. The most likely outcome can be related to the relative degeneracy of the three distributions. There are two ways of arranging 2 objects, A and B, in 2 boxes, namely object A in box 1 and B in box 2 or *vice versa*. There is only one way of arranging 2 objects in 1 box, namely both A and B in box 1. Thus the solution 1 H and 1 T is twice as likely as the solution 2 H or 2 T. It has twice the *degeneracy*. Entropy[1] is a measure of this degeneracy. In this chapter we will use logarithms to the base 10 rather than base 2: this is so that the reader can easily follow the calculations since most

[1]defined by $S = - \sum\limits_{n=1}^{n=N} p_n \log_2 p_n$: see Eq. (4.30).

calculators and spreadsheets compute logarithms to the base 10 directly, but it makes no difference to the relative entropies of various distributions. We refer to such entropies as S_{10} rather than S.

Table 5.5 - Entropy and degeneracy of placing five objects in two boxes

State	Entropy	Degeneracy
5 H	0	1
4 H 1 T	0.217	5
3 H 2 T	0.292	10
2 H 3 T	0.292	10
1 H 4 T	0.217	5
5 T	0	1

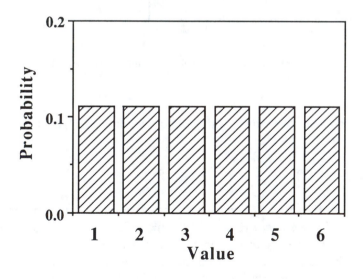

Fig. 5.18 - Most likely frequency distribution for an unbiassed six sided die.

The entropy for 1 H and 1 T is $-\log_{10}2 = 0.301$, whereas that for 2 H or for 2 T is $-\log_{10}1 = 0$. The distribution with the maximum entropy is the the most likely distribution. These entropies are not exactly equivalent to the relative degeneracies, but it can be shown that for an infinite number of objects and states, they converge to probabilities. The entropy and related degeneracies for placing five objects in two boxes (i.e. the possible distributions for tossing five coins) are tabulated in Table 5.5. The ratio of degeneracies is approximately similar to the ratio of entropies. This problem is analogous to that of a spectrum in which we observe only at two points but one in which the spectrum is very coarsely digitized in the y direction, in other words there are only five possible detection levels, so that readings of intensity can only take an integral value between 0 and 4. In practice we can measure the intensity of peaks much more accurately, and the relative entropy of any possible distribution is given by the continuous form (Eq. (4.30)).

In the examples given above we assume very little about the system, apart from a lack of bias. The most likely distribution (that of *maximum entropy*) is the flattest distribution. If we have no preconceived knowledge about a spectrum and have performed no experiments previously then this distribution is the most likely: any other distribution assumes features for which we have no experimental evidence. A slightly more realistic situation can be analysed by considering the case of a biassed die, (see Jaynes' book). Consider an experiment in which a six sided die is tossed several times and it is established that the mean reading is 4.5, rather than the normally expected reading of 3.5. We want to predict the underlying distribution of readings. This problem is typical of chemical experiments – we normally obtain some information and want to use this to predict a fuller model of the system. For an unbiassed die the maximum entropy solution is a flat distribution (Fig. 5.18), but this is not possible for a biassed die, since the mean of the distribution, given by

$$\mu \quad = \quad \sum_{n=1}^{n=6} n \, p_n \qquad\qquad (5.28)$$

where n is the value of the die (between 1 and 6), is no longer 3.5. There are, of course, an infinite number of distributions that provide a mean of 4.5, but which is the *most likely* distribution in the absence of further knowledge or experimentation? We illustrate three possible distributions in Fig. 5.19. In the top two distributions some possible states have zero probability. This is somewhat unlikely in the absence of further information. The third distribution is of the form

$$p_n \quad = \quad 0.0238 + 0.0571 \, n \qquad\qquad (5.29)$$

(all probabilities are non-zero and the probability of obtaining successfully increasing values of the die increases linearly). The entropy of the three distributions pictured in Fig. 5.19 increases from 0.301 to 0.602 to 0.693 as the interested reader should be able to verify, suggesting that entropy is a good measure of increasing likelihood. Of course we might have some further information about how the die is biassed. For example, one face of the die might be favoured more than the rest which are equally favoured. Alternatively the bias might involve a weight being placed in the die, so that the faces closest to the weight are increasingly favoured, according to the distance from the weight: we would have to know the physical distribution of the faces of the die. But these possibilities involve extra *knowledge* of the system, and in the absence of such knowledge we must construct a simpler model. Below we will discuss how elaborate models and knowledge of data can be incorporated into entropy calculations.

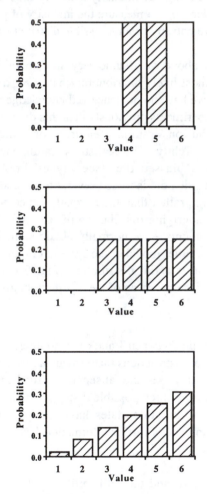

Fig. 5.19 - Distributions of a biassed die (μ=4.5) of increasing entropy (from top to bottom).

The problem of the biassed die resembles a simple chemical problem, in which six spectroscopic or chromatographic intensities are measured. Some partial knowledge of a system is available and from the partial knowledge we want to deduce a better model. The theme of this chapter is deconvolution or deblurring of analytical signals in order to use chemometric methods to improve the quality of information available from spectra or chromatograms. As discussed in Section 5.3, convolution normally involves changing peakshapes, often from flat broad peaks to sharper better resolved peaks. Often we can choose a model for the underlying peakshape. If the deblurred peakshape is given by $h(\zeta)$, then, we assume that a blurring function $s(\zeta)$ contributes to the observed peakshape, so that

$$f(\zeta) \quad = \quad h(\zeta) \otimes s(\zeta) + v \qquad\qquad (5.30)$$

where $f(\zeta)$ is the observed data and v a noise function. This equation is similar to Eq. (5.24), but in this case the deblurred peakshape is unknown. One way of determining $h(\zeta)$ from the observed spectrum $f(\zeta)$ is to test various models of the unblurred peakshapes, and see how well these models fit the observed data. Most chemometric methods involve reducing noise rather than reducing the blurring process. It is easy to propose a noise free model by approaches such as minimizing least squares of the residuals between the model and the observed points, but less easy to determine a deblurred model. Consider a simple example of a blurred peak which appears roughly rectangular but slightly asymmetric (Table 5.6). Perhaps we want to compare two possible models for the deblurred peakshape. The first is one in which the peak actually consists of two symmetric triangular peaks. The intensities for this model are given in the table. A second possibility is that the peak is roughly triangular, but slightly asymmetric. We constrain both models so that the net intensity remains constant. As seen from Table 5.6, both these proposed models have a residual sum of squares

$$\sum_{i=1}^{i=6}(x_i - \hat{x}_i)^2 \quad = \quad 1.5 \qquad\qquad (5.31)$$

so there is nothing to choose between them using least squares criteria. However, the *entropy* of the two models[1] differs: the solution with two symmetric triangular peaks is the one with the highest entropy and so more likely. Note that the original data have a slightly higher entropy still to either of the solutions, but the entropy criterion is not the only criterion used to pick an optimum model, otherwise we would always choose a flat spectrum. We must take into account the raw data (hence the least squares fit) and the sort of model we want. Of course, we could formalize the type of model we test, restricting it to symmetric triangles, for example, or sharp "spikes". This is our choice.

---

[1] Remember that the probabilities must be normalized i.e. divided by 8 in this particular example.

Table 5.6 - Examples of how maximum entropy criteria can be employed to choose between two models with identical least squares fit

i	x_i	$\hat{x}_{i(1)}$	$\hat{x}_{i(2)}$
1	1.5	1.0	1.0
2	1.5	2.0	1.5
3	1.5	1.0	2.5
4	1.5	1.0	1.5
5	1.5	2.0	1.5
6	0.5	1.0	0.0
$\sum_{i=1}^{6}(x_i-\hat{x}_i)^2$		1.5	1.5
$-\sum_{i=1}^{6}p_i\log_{10}p_i$		0.753	0.680

Often least squares criteria move the solution in the opposite direction to entropy criteria, so both approaches are complementary and can be employed effectively to reach an optimal model. One way of using entropy criteria in model building is to compare solutions with identical least squares errors and choose the solution with maximum entropy.

One of the most sophisticated applications of maximum entropy is in Fourier transform NMR spectroscopy. The reader is referred to Section 5.2 for a brief description of Fourier transform terminology. In this approach data are acquired in the *time domain* for reasons of efficiency (as explained in Section 4.7). It is desired to produce a noise free model of the frequency domain. A linear approach might be to Fourier transform the time domain and then use a smoothing function such as the Savitsky-Golay method, but as discussed above this is fairly limited when there is sophisticated knowledge of lineshapes and / or noise. The NMR implementation is as follows.

(1) A noise free frequency domain is guessed. This consists of a sum of several peaks of a given shape (e.g. Lorentzian as defined in Section 5.6). Often a flat map is used in which it is assumed that there is a single peak of equal intensity at each given frequency in the spectrum (which might be observed at 1000 sampling points).

(2) This noise free frequency domain is inverse transformed to an estimated noise free time domain $(\hat{x}_i)$ which is the first guess to the real noise free time domain.

(3) The sum of squares of the residuals between this guessed time domain and the observed time domain (x_i) is computed. Normally a test such as the χ^2 test is used to test how small these residuals are. If they are sufficiently small the model is said to have converged. Otherwise, step 4 is followed.

(4) The value of the χ^2 (or of the residual sum of squares or similar statistic) is then reduced by considering $\partial\chi^2/\partial x_i$ (or relevant statistic or residual sum of squares in the time domain): a new frequency domain is found that reduces the time domain residual sum of squares by a given amount. There will be a large number of solutions for a given value of χ^2 or residual sum of squares. The solution with *maximum entropy* is chosen and this is the next guess. Step 3 is then repeated.

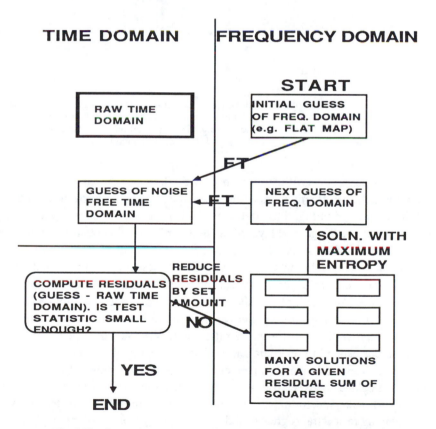

Fig. 5.20 - Computational implementation of maximum entropy in Fourier transform spectroscopy.

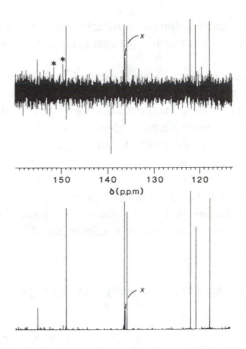

Fig. 5.21 - Result of maximum entropy reconstruction (below) of a noisy
NMR spectrum. Instrumental artifacts are indicated by an asterisk or star.
The starred artifacts are removed during the deconvolution. Reprinted by
permission from *Nature* Vol.311, p. 446 © MacMillan Magazines, Ltd.

This approach is illustrated in Fig. 5.20. A peakshape model can be
incorporated: the effectiveness of the maximum entropy approach in solution
NMR is indicated in Fig. 5.21. In the original spectrum, there are several sharp
datapoints above the general level of the noise and it would not at first glance be
certain which of these corresponds to real peaks and which to noise. The
maximum entropy solution establishes which of these are real peaks and which
are not. Independent experiments can be used to verify that this solution is
correct in the case cited: obviously in many practical cases it is not so easy to
directly verify the solution. However, maximum entropy has enabled us to
reliably improve the signal to noise ratio and determine which peaks are most
likely to correspond to real data and which are least likely. This approach has
immense advantages and is complementary to signal averaging. Obviously there
must be a balance between use of computational approaches such as maximum
entropy to make spectra more informative and building better instruments or
averaging more spectra. A great deal depends on the problems of instrument time
and chemical stability. If a compound is stable and a reasonably good spectrum
can be acquired in 10 s, it might be worth acquiring 100 spectra, over 1000 s (or

16 min) to improve the signal to noise ratio 10 times: this is often the case in routine ^{1}H NMR, for example. In ^{13}C NMR of very small quantities of compound it may take several hours to record a spectrum of adequate signal to noise ratio. Under such circumstances it is worth investing time to use maximum entropy methods to improve the spectrum still further. With modern developments in computing, the maximum entropy approach, which is computationally intensive, becomes more attractive, whereas the first large applications were performed off-line on relatively slow mainframe computers, so computing time was more expensive relative to instrument time. It is vital to assess the problem and what resources are at hand, and consider the balance between instrument time, importance of the problem and computational resources before deciding whether to use maximum entropy approaches. Despite these limitations, the maximum entropy method has been very effective in many well-publicized cases.

Maximum entropy non-linear deconvolution has been applied to a large number of chemically useful techniques such as Raman, IR, neutron diffraction, X-ray diffraction etc. The method has even broader applications in image processing. In fact a spectrum (or chromatogram) could be regarded as a blurred image, and the principles of image enhancement are the same.

Much debate centres on criteria for entropy. The original definition by Shannon (Eq. (4.30)) will, in the absence of other information, result in a maximum entropy solution that is a flat distribution. Obviously the experimental evidence changes the solution, but only because least squares or equivalent criteria take priority over the maximum entropy solution. Sometimes there is more specific *prior knowledge* of the solution. This knowledge can be incorporated into the maximum entropy criterion by using a prior model, so that

$$S_{10} = -\sum_{n=1}^{n=N} p_n \log_{10}(p_n/m_n) \qquad (5.32)$$

In such a case, the solution with maximum entropy will be the one that exactly obeys this model. The experimental observations may suggest deviations from this model, and methods such as iterative minimizing least squares fit of the predicted model to the observed data will produce a different solution. But the preferred solution in the absence of other information will be the model specified by m_n rather than a flat map. This approach has many advantages, and allows the chemist to incorporate his own thoughts and ideas prior to using maximum entropy techniques.

Several other definitions of entropy will be found in the literature, one of the most famous being that due to Burg, which specifies entropy as

$$S_{10} = +\sum_{n=1}^{n=N} \log_{10} p_n \qquad (5.33)$$

This approach has mainly been used in time series analysis.

The user of maximum entropy approaches must consider very carefully *how much he knows about the system in advance* and use this information in the model or definition of entropy. There is no point applying a mathematically sophisticated approach if simple intuitive information needs to be discarded in order to use the method. Fortunately maximum entropy can be employed as a very flexible method of non-linear deconvolution. Constraints such as all peaks must be positive are very easy to incorporate (in fact the normal method assumes positivity), and more subtle information such as lineshapes can also be put to use. Such methods differ from approaches to linear deconvolution in which constraints and physical knowledge are usually impossible to incorporate. However, the choice of method must, as always, depend on how much is known about the system and what information is required from the data. There is no universal solution and the chemometrician must always look at problems as a whole. Maximum entropy is one tool that the chemometrician should consider if he has expensive problems with significantly low signal to noise ratios that cannot easily be improved by instrumental methods.

It should be noted that there are, also, several other approaches to non-linear deconvolution. Many of these were developed before the advent of fast computing power so are less expensive on resources than maximum entropy. However, they are normally fairly specialized in that they have only been applied to one analytical techniques and do not approach the universality of the maximum entropy method. One of the most widespread methods for non-linear deconvolution is *Jansson's method* (see bibliography) which has been applied in optical spectroscopy. It would be impossible in a text this size to outline all the approaches to non-linear deconvolution in detail but the chemometrician should always explore the possibility of employing such methods.

5.5 DERIVATIVES

A common form of resolution enhancement involves using *derivatives*. Strictly speaking these are a form of linear deconvolution, but, historically, they are often considered a separate technique: this is probably because it is easy to build instruments that record derivative spectra directly without any computational processing, so some methods of spectroscopy, most notably electron spin resonance (esr) spectroscopy are conventionally recorded in the derivative mode.

As increasingly higher derivatives are computed, partially overlapping peaks should become resolved (Fig. 5.22). This follows from simple intuitive assumptions about the nature of pure peaks: a pure peak should only have one inflection point (normally a maximum). The spectrum at the left of Fig. 5.23 is normally considered to arise from a pure peak, whereas the spectrum at the right represents a cluster of at least two partially overlapping peaks. Of course there is no physical reason why a pure peak should not have multiple inflection points, but this is most unlikely. This extremely simple assumption about peakshapes explains the effectiveness of derivative techniques.

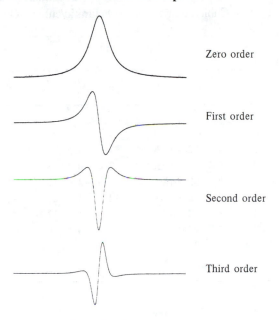

Zero order

First order

Second order

Third order

Fig. 5.22 - Use of derivatives.

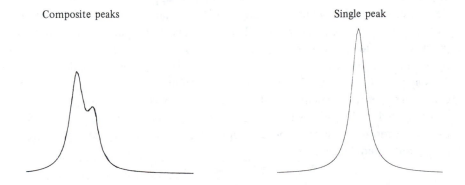

Composite peaks

Single peak

Fig. 5.23 - Simple knowledge of peakshapes: the peak on the right is likely to arise from more than one component whereas that on the left is consistent with a single component.

Table 5.7 - First and second derivative Savitsky-Golay coefficients

	First derivative						Second derivative					
	Quadratic			Cubic/quartic			Quadratic/cubic			Quartic/quintic		
	No. Points											
	5	7	9	5	7	9	5	7	9	5	7	9
j												
−4	-	-	−4	-	-	86	-	-	28	-	-	−4158
−3	-	−3	−3	-	22	−142	-	5	7	-	−117	12243
−2	−2	−2	−2	1	−67	−193	2	0	−8	−3	603	4983
−1	−1	−1	−1	−8	−58	−126	−1	−3	−17	48	−171	−6963
0	0	0	0	0	0	0	−2	−4	−20	−90	−630	−12210
+1	1	1	1	8	58	126	−1	−3	−17	48	−171	−6963
+2	2	2	2	−1	67	193	2	0	−8	−3	603	4983
+3	-	3	3	-	−22	142	-	5	7	-	−117	12243
+4	-	-	4	-	-	−86	-	-	28	-	-	−4158
Normalization	10	28	60	12	252	1188	7	42	462	3	99	4719

Note : To compute the coefficients divide by the normalization constant.

One of the problems of taking derivatives, though, is that noise is amplified very considerably: normally noise is much less smooth than peakshapes, which explains the effectiveness of moving average approaches. In the case of derivative spectroscopy this works to our disadvantage. Theoretically any non-overlapping peaks can be resolved using derivative spectroscopy, but, in practice, the effectiveness of the method is limited by the level of noise. This can be overcome computationally by simultaneous smoothing and taking derivatives. This approach is a form of linear deconvolution and moving average methods can be used for rapid calculation of derivatives. Savitsky-Golay coefficients can be computed in the same way as in Section 5.3, as are tabulated in Table 5.7. Literature references extend these computations, if necessary, to higher order polynomials and datalengths.

In addition to resolving peaks, derivatives can be used for accurate estimation of positions of peaks. The centre of a first derivative peak should cross the baseline so making it easy to pinpoint the exact position of a peak. If a peak is fairly flat, the exact centre can be hard to estimate in the zero order spectrum. The disadvantage of derivatives is that information about integrals is lost. However, frequently the experimenter is only interested in how many peaks are in a cluster and what their positions are, for which derivatives represent a useful approach.

The use of and assumptions in the calculation of derivatives is vastly less sophisticated than for non-linear deconvolution methods, and yet these conceptually simple approaches are, actually, frequently employed in chemistry. The advantage is that extremely simple assumptions are made about the data, so the chemist does not need elaborate knowledge of the system. Also derivatives are easy to compute. However, if the chemometrician does have a relatively sophisticated model of the signals or the noise, then he should put this to use, and under such circumstances more elaborate approaches are more appropriate. The chemometrician should estimate how much time is worth spending on a problem. Building sophisticated models can take much time and experimentation; using approaches such as the maximum entropy approach involves expertise and considerable computational sophistication for relatively large problems. If, though, a system is to be repeatedly studied it may be well worth modelling the system in detail: for example if a particular instrumental method is to be used everyday and good quantitative data are required of the system, then it may be appropriate to invest several months investigating the nature of the data and the associated problem of noise and modelling peakshapes. On the other hand, approaches such as derivative spectroscopy are good examples of simple and effective methods that provide powerful, but rather limited, insight into a dataset.

5.6 LINESHAPE ANALYSIS AND CURVE FITTING

Information about lineshapes can be used to advantage in signal processing. Some of the methods listed above cannot incorporate detailed models for lineshapes: these include the use of derivatives which assume only that pure peaks have one inflection point (normally a maximum) and use of Savitsky-Golay moving averages which make assumptions only that the noise is less correlated than signals. Other approaches such as some of the time domain methods for linear deconvolution can easily incorporate sophisticated lineshape models. Non-linear methods such as maximum entropy approaches can incorporate almost indefinite levels of knowledge about the system, but the model must be very explicitly stated and default use of such approaches will not incorporate complex knowledge. A further set of methods, often called *curve fitting* or *constrained minimization* where a series of peaks are fitted to detailed peakshape models, will be discussed in this section.

The chemometrician often has some information as to the shape of a peak. Two common shapes are the Gaussian where

$$f(\zeta) \quad = \quad I_0 e^{-(\zeta-\zeta_0)^2/\Delta^2} \tag{5.34}$$

where I_0 is the height of the peak, ζ_0 is the position of the centre of the peak and Δ is the related to the width at half height. It is easy to show, by simple mathematics, that

$$\Delta_{1/2} \quad = \quad 2\Delta\sqrt{(\ln 2)} \tag{5.35}$$

where $\Delta_{1/2}$ is the width at half height, so that Δ is directly proportional to the half width. These parameters are illustrated in Fig. 3.1. This formula for a Gaussian should be compared to the normalized statistical formula given in Eq. (3.1), and differs only in various scaling constants. Gaussian lineshapes commonly occur in, for example, uv / visible and IR spectroscopy. Another common distribution is a *Lorentzian*, often characteristic of solution NMR spectroscopy, given by

$$f(\zeta) \quad = \quad \frac{I_0}{(1+(\zeta-\zeta_0)^2/\Delta^2)} \tag{5.36}$$

The three parameters have the same significance as for the Gaussian, but the relation between Δ and the width at half height is now slightly different and is given by

$$\Delta_{1/2} = \quad 2\Delta \tag{5.37}$$

as can be verified. These parameters are illustrated in Fig. 5.24. This peakshape is also related to a *Cauchy distribution*. Obviously the formula for both types of peakshape can be scaled and normalized as desired, so it is important to check conventions when reading papers. In some cases such as MS and chromatography peakshapes may be *skew* or asymmetric. There are a number of ways of modelling these peakshapes. Some investigators have tried to produce exact models and an exact understanding of these peakshapes based on physical and instrumental processes. However, such approaches are not really within the realms of chemometrics, and belong more properly to physical chemistry. In Chapter 1 we discussed the difference in philosophy between traditional chemistry in which considerable effort is spent to produce exact models of a system and the more statistical chemometric approach. Normally an empirical, statistical, model suffices since experimental data are not always sufficiently accurate to allow for more detailed understanding of the process. In fact if too much effort is placed in deriving peakshape models, the chemist may well become diverted from the main problem which is the optimization and interpretation of chemical experiments: analytical techniques are means to an end rather than ends in themselves. It is often difficult to keep a balance between understanding a system (spectroscopic or chromatographic) in detail and using more statistical approaches which may not result in quite such a good understanding of the underlying processes, but may well reach a more useful answer to the analytical problem faster. The chemist normally errs on the side of trying to understand the system in detail. This psychological inclination of chemical researchers and the resistance that chemists have towards employing a statistical approach has already been emphasized in earlier chapters. Obviously though, simple and well established information about a system, such as knowing that peakshapes are reproducibly Gaussian or Lorentzian should be employed where necessary.

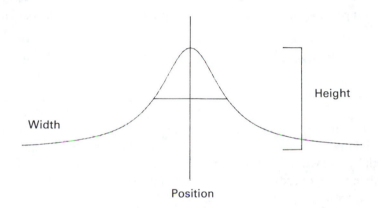

Fig. 5.24 - Parameters that characterize a Lorentzian or Cauchy distribution.

One of the simplest forms of deconvolution is to use non-linear methods in the time domain.[1] It is easiest to understand how the choice of deconvolution function depends on peakshape by considering the concept of *Fourier pairs*. Every function in the frequency domain corresponds to a unique function in the time domain. It can be shown that the Fourier transform of a Gaussian as given above gives another Gaussian of the form

$$f(t) \quad \propto \quad I_0 \Delta e^{-t^2 \Delta/4} \cos(\zeta_0 t) \tag{5.38}$$

where the constant of proportionality depends on the normalization constant in the integral, and whether radians or cycles s^{-1} are chosen. The Fourier transform of the Lorentzian gives a decaying exponential of the form

---

[1] As discussed in Section 5.3 (Eq. 5.25) moving averages in the frequency domain are simply equivalent to multiplication.

$$f(t) \quad \propto \quad I_0 \Delta e^{-|t|\Delta} \cos(\zeta_0 t) \tag{5.39}$$

The reader is should be able to verify the correspondence between frequency and time domain parameters as listed in Table 4.15; for example, the decay rate is faster the broader the peaks. The important conclusion of this section, though, is that, from knowing the shape of a function in one domain it is possible to predict the shape in the other domain. Some Fourier pairs are tabulated (Table 5.8).[1]

Table 5.8 - Fourier pairs

Frequency domain	Real time domain
Delta function at position ζ_0	Cosine wave of frequency ζ_0
Lorentzian at position ζ_0	Decaying exponential imposed on cosine wave of frequency ζ_0
Gaussian at position ζ_0	Decaying Gaussian imposed on cosine wave of frequency ζ_0

This information helps devise simple linear deconvolution functions. For example, a Lorentzian of width Δ in the frequency domain, corresponds to an exponentially decaying function in the time domain. In order to reduce the peakwidth in the frequency domain, it is merely necessary to multiply this by a positively increasing exponential, so that

$$f'(t) \quad = \quad e^{+\lambda t} f(t) \tag{5.40}$$

where λ is a positive number. The line width in the frequency domain now is reduced by the ratio $(\Delta - \lambda) / \Delta$. The effect of varying the value of λ is illustrated in Fig. 5.16 for a pair of peaks, and it can be seen that resolution increases with increase in λ. The sharpest possible peak will be obtained when $\lambda = \Delta$. Clearly, if we have some knowledge of the peak widths in advance we can plan the optimal strategy for filtering. This approach, however, seems too good to be true. Can we resolve any overlapping peaks providing they do not coincide if we know the lineshapes in advance? Commonsense suggests that this is not possible.

A crucial factor that we omitted in discussion above is that there will always be *noise* in the measurement process, so the amount of information is limited. A positive exponential will emphasize noise as well as resolution, and, therefore, the advantage of the filter function will be negated by the increase in noise.

---

[1] A delta function is a sharp spike one point in width (or infinitely sharp in the continuous case).

Clearly the optimal solution will balance the increase in resolution against the increase in noise. One approach is to modify the function so that

$$f'(t) \quad = \quad e^{+\lambda t - \mu t^2} f(t) \tag{5.41}$$

where λ and μ are positive numbers. The term in μ is more gentle and reduces the noise at the end of the time series. If the exact width of the line is known in advance it is possible to devise a *matched filter* whereby $\lambda = \Delta$. Under such circumstances the lineshape in the filtered time series will be determined by the value of μ and will be a Gaussian. The width can be controlled and the choice of μ is normally determined by signal to noise factors. This filter is described by some workers as an *optimal filter* and is frequently used in NMR spectroscopy. However extreme care must be taken when using so-called optimal filters. It is certainly possible to derive theoretically appropriate expressions for a given lineshape, signal to noise ratio and so on, but these depend on detailed knowledge of the system. Clearly if we know everything about a system, there is no point in further dataprocessing, or chemometrics, so we are normally working with *partial knowledge* of a system. If misapplied partial knowledge can be very dangerous. For example an optimal filter devised for a Lorentzian lineshape is not normally appropriate for a Gaussian lineshape. If there is limited knowledge available of a system it is probably best to use simpler approaches such as unweighted moving averages and derivatives as described in Sections 5.3 and 5.5. Alternatively non-linear methods such as the maximum entropy approach can be powerful. Simple knowledge such as "all peaks have positive intensities" can be very easily coded into non-linear approaches for signal processing.

If there is strong knowledge of lineshapes, it is also possible to code this into non-linear approaches, but sometimes quite hard. Eq. (5.32) allows the incorporation of a model into the definition of entropy, but careful thought is frequently necessary prior to using models. If there is firm knowledge of lineshapes, maximum entropy approaches may not always be very successful. This is simply because so much is known about the system anyway, that there is little point applying methods that have been developed for data for which there is very limited analytical knowledge.

What to do with peakshape information is typical of the central dilemma of chemometrics. In Chapter 1 we discussed the problem of experimental errors and explained that in traditional branches of physical chemistry, experimental data are better than models, so data analysis consists of developing better models and problems of experimental errors are, effectively, ignored. Chemometrics helps when experimental errors are significant: in the context of signal processing this means slightly noisy experiments or situations where there is incomplete knowledge of, or data about, the system. Many spectroscopists resist chemometric approaches because they are used to working with and developing exact models and are highly suspicious of experiments where they cannot fully understand the physical basis of the data: the experimental physical scientist will often abandon experiments if he is unable to obtain exact understanding of the underlying process, or else will invest substantial and often unnecessary time and effort studying the physical basis of the process. However, the desired result of a spectroscopic experiment might be in terms of the structure of a compound or the proportion of a compound in a mixture and not a set of equations that explain the physical basis of a highly complicated lineshape. Rather than discard experiments for which there is inadequate information to completely model the data, it is important to be able to work with partial knowledge of a system and to be able to make reasoned assumptions as to what sort of information can be reliably deduced from the data. In this way, the chemometrician uses time more efficiently and concentrates on solving practical problems even if he has insufficient time and resources to adequately understand molecular processes in detail.

Often quantitative information is required from spectroscopic and chromatographic measurements. This can be estimated from deconvoluted data providing some care is taken. For example, if all peaks have identical shapes and widths it is quite easy to design filters that preserve relative integrals. If peaks have different widths, most filters will distort this information. Sometimes this matters a great deal in the resultant interpretation and there will be serious errors in the quantitation of various peaks. In other cases this may not be very serious. In Chapter 7 we discuss techniques for preprocessing data prior to the use of multivariate techniques. If, for example, a dataset consists of 20 peaks, measured for a set of 10 spectra, and if each of the 10 sets of the peaks is weighted so that they have constant mean and standard deviation,[1] it does not matter if that there is an absolute but consistent error in the estimation of the relative measurement of a given peak, providing the error is the same throughout the series of spectra. On the other hand, if absolute quantitation is required this error could seriously distort the resultant interpretation of the data. Therefore, chemometricians must take into account how signals have been processed prior to employing multivariate and related techniques for interpretation of chemical data.

A different approach to the quantitative interpretation of chromatograms and spectra involves *curve fitting*. The basis of this method is to minimize the residual sum of squares between the observed data and a peakshape model. This approach is sometimes also called *constrained minimization*. There are a huge number of peakshape models in the literature, often differing according to the particular analytical technique, but there is insufficient room in this text to comment on the merits of various models in detail. However, somewhere a line

[1]This is called standardization.

must be drawn between chemometric approaches and traditional physical chemistry. Exact physical modelling of peakshapes is not really a part of chemometrics but belongs to the vast and well established domains of physical and analytical chemistry. On the other hand chemometrics can help comparing simple models of peakshapes.

Normally a constraint is placed on the peakshape model. For example, we might wish to know whether a cluster of peaks consists of two or of three overlapping Gaussians. A mean peakwidth might already be known from independent experiments. If the area of the cluster is normalized,[1] an N peak model consists of $2N-1$ parameters namely

(1) The positions of N peaks

(2) The relative areas of $N-1$ peaks (remembering that the overall area is normalized).

Standard software for minimizing residuals in $2N-1$ dimensional space can then be used.[2] The chemometrician might want to compare the two peak model to the three peak model. Obviously the residual sum of squares for the three peak model will be less than that of the two peak model, but the chemometrician needs to estimate the *significance* of the extra terms. Traditionally, spectroscopists use tests such as the χ^2 test to compare the observed (raw data) to the model: this test assumes that errors are normally distributed. It is possible to determine a target value of χ^2: the value of N for the target χ^2 is usually taken to be the number of peaks in the cluster. Adding extra peaks to the model will always improve goodness-of-fit but the significance of this improvement is small. In Section 2.5, we discussed an alternative approach to assessing the magnitude of errors when increasing the complexity of a model which was to estimate the analytical error (due to noise) and then use ANOVA to compare the residual lack-of-fit to the model: this approach involves estimating the magnitude of the noise as well as how well the peakshape model has been obeyed, but provides far more insight into the problem. Spectroscopists and chromatographers rarely do this, probably because most have been schooled by numerate chemists rather than statisticians.

It is possible to become confused as to the optimum use and interpretation of curve fitting methods in chemometrics. Sometimes these methods can result in several close minima, and very small changes in assumptions, iteration rates and so on can yield radically different solutions. Under such circumstances it is often best to change the approach altogether and use non-linear methods for deconvolution, such as maximum entropy which can function well on limited

---

[1] This transformation normally involves recalculating intensities so that the overall intensity is a constant (e.g. 1) and is useful for comparing peakshapes.

[2] There are a large number of algorithms but the comparison of the merits of various approaches is outside the scope of this text. Nearly every numerical subroutine library and statistical package has a large variety of methods available.

knowledge of peakshapes. There is no universal approach and the technique employed must depend on the unique problem under consideration.

Finally the nature and knowledge of noise distributions is another consideration that should be taken into account when choosing the method for signal processing. Unfortunately many chemists are reluctant to invest time and effort to directly study the nature of noise in a system, despite the fact that this information can help solving problems. In many chemical situations noise is *white non-stationary,* that is the noise at each successive sampling point is unrelated to the noise at previous or prior sampling points, but the overall noise can be described by a distribution such as a uniform distribution (see Section 4.2) or a Gaussian distribution. Under such circumstances the techniques described in this chapter are most valuable. In other cases noise is *coloured* and can usually be described by ARMA processes (see Section 3.2), where each successive value of the noise is partially dependent on previous values of the noise and different approaches to signal enhancement may be of value. *Time series analysis* involves Fourier transforming a correlogram (see Section 3.2 for a discussion of correlograms). Because this type of noise distribution is rarely reported in spectroscopy or chromatography we will not discuss these methods in this chapter, but the reader is referred to Section 7.8 where time series methods are discussed in the context of pattern recognition. Such approaches are not very widely employed in chemical signal processing, probably because most spectroscopists and chromatographers have concerned themselves principally with direct study of signals rather than noise. With the advent of chemometric thinking, this attitude may change. The chemometrician wants to deduce as much information, as efficiently as possible, from a system, and in many cases it is worth investing time studying the nature of noise in the system in order to reach a better interpretation of experimental data.

5.7 FAST ON-LINE FILTERING BY THE KALMAN FILTER

Historically, much interest has centred on fast on-line filters. We have not discussed these in detail above, concentrating on off-line data-processing. For completion we include a brief section on a method called the Kalman filter. The modern day experimentalist has available cheap fast computing power, extensive disc storage and powerful second processors so can normally analyse data at leisure. However several years ago, most data was recorded on-line and there were rarely chances to return later and examine trends in the data. It was, therefore, important to devise rapid methods for smoothing and modelling signals in real time. The Savitsky-Golay approach is one method for rapid filtering, but rather limited knowledge of the data is incorporated into the model and it requires information from several datapoints for each estimate – in order to provide a smoothed estimate for x_i, $2N+1$ values of x need to be known.

A much more general form of rapid filter is called the *Kalman filter*. This approach originated in electrical engineering, where the characteristic behaviour of electric circuits is often well established. Reduction of noise can result in sharper pictures, better sound and so on. Many early spectroscopists and chromatographers were instrument builders and computer hardware specialists so the Kalman filter was a natural area of interest.

The objective of rapid on-line filters is to use the past history of a sequential series to predict the future trends and then compare the observed with predicted readings so as to reach a smoother estimate of a system. If x_i is the observed value of a parameter recorded at point i, then this parameter can often be improved upon by using a model based on the previous observations in the series. This is somewhat analogous to some of the spatial sampling strategies in geochemistry described in Section 3.3. Although we can sample any point directly, it is more accurate to use data from neighbouring points, and estimate the spatial change in the value of parameter of interest (e.g. by a variogram) and then re-estimate the value of the parameter at the particular sampling point from this information: most underlying temporal and spatial processes are fairly smooth in contrast to noise distributions so this method often reaches a more accurate estimate of the signal than direct measurement.

To understand the Kalman filter we consider a process in recorded in time, where x_i is the ith value of a parameter x. A *prior* estimate of x_i can be obtained from the previously measured value x_{i-1} and a model of the processes that cause variation in the time series before we actually record the experimental datapoint: we call this predicted value $\hat{x}_{i|i-1}$. However, once x_i has been recorded we can refine this prediction by comparing the observed value (which will be subject to experimental error) to the *prior* estimate. The new value is called $\hat{x}_{i|i}$. The Kalman filter equation computes this new estimate by

$$\hat{x}_{i|i} \quad = \quad \hat{x}_{i|i-1} + K_i(x_i - \hat{x}_{i|i-1}) \tag{5.42}$$

The parameter K_i will be discussed below. At the same time the Kalman filter also provides estimates of the error in $\hat{x}$. We call these mean square errors $\hat{v}_{i|i-1}$ and $\hat{v}_{i|i}$ corresponding to the predicted error before and after recording the value x_i. Two other pieces of information need to be determined. The first is the mean square instrumental error given by r (which is usually assumed to be constant throughout the measuring process) and the second is an ideal model for how the system is expected to change with time given by

$$\hat{x}_{i|i-1} \quad = \quad f(i,i-1)\,\hat{x}_{i-1|i-1} \tag{5.43}$$

The functional dependence must be estimated from knowledge of the system. Sometimes there is no particular knowledge so this function [$f(i,i-1)$] is merely set at 1.

The value $\hat{v}_{i|i-1}$ predicted prior to recording x_i is given by

$$\hat{v}_{i|i-1} = [f(i,i-1)]^2 \, \hat{v}_{i-1|i-1} \tag{5.44}$$

The new estimate is given by

$$\hat{v}_{i|i} = \hat{v}_{i|i-1}(1-K_i) \tag{5.45}$$

A final equation allows us to compute K_i by

$$K_i = \frac{\hat{v}_{i|i-1}}{\hat{v}_{i|i-1}+r} \tag{5.46}$$

Table 5.9 - Kalman filter example

i	x_i	$\hat{x}_{i-1|i}$	$\hat{v}_{i-1|i}$	K_i	$\hat{x}_{i|i}$	$\hat{v}_{i|i}$
1	1.1150	1.0000	0.0400	0.7937	1.0913	0.0083
2	0.5822	0.8934	0.0055	0.3472	0.7854	0.0036
3	0.8793	0.6430	0.0024	0.1888	0.6876	0.0020
4	0.6598	0.5629	0.0013	0.1123	0.5738	0.0012
5	0.6588	0.4698	0.0008	0.0700	0.4830	0.0007
6	0.1169	0.3954	0.0005	0.0448	0.3830	0.0005
7	0.3112	0.3135	0.0003	0.0292	0.3134	0.0003
8	0.3516	0.2566	0.0002	0.0192	0.2585	0.0002

The value of K_i must be between 0 and 1 in value. Sometimes K_i is held constant throughout the filtering process, but it can be iteratively adjusted if required. Generally, though, if K is varied throughout the measurement process it decreases in size: this is because the future estimates, $\hat{x}_{i|i-1}$, of the model become progressively better throughout the experiment as more is known about the system. The use of Kalman filters is best illustrated by a numerical example. Table 5.9 lists the values of a process that is believed to be approximately exponential, with a decay rate of $e^{-0.2i}$: therefore $f(i,i-1)$ remains constant throughout the process at 0.8187 ($=e^{-0.2}$). The system noise is believed to have a mean square value of 0.0104.[1] Initial guesses are required for $\hat{x}_{0|1}$ ($=1$) and $\hat{v}_{0|1}$ ($= 0.04 = 0.2^2$). The smoothed estimates $\hat{x}_{i|i}$ are illustrated in Fig. 5.25.

The Kalman filter approach can be refined to incorporate very sophisticated models or simple polynomial smoothing functions. A central theme of this chapter is appreciating how much or little we safely know about a system and

---

[1] The interested reader can show that this is the expected value for a uniform distribution between 0.5 and -0.5 : see Eq. 4.2.

some thought must be given to the form of model and estimates of noise. Kalman filters probably are most useful for fairly repetitive measurement procedures, and can readily be incorporated into on-line software. These situations only rarely occur in chemometric situations although are common in physical chemical experiments such as reaction kinetics or spectroscopy. Although Kalman filters are certainly useful aids it is wondered whether this approach is more suitably classed as a computational method used by chemists rather than as a chemometric tool. The approach has certainly though been shown to be an effective tool and there is much valuable literature in this area.

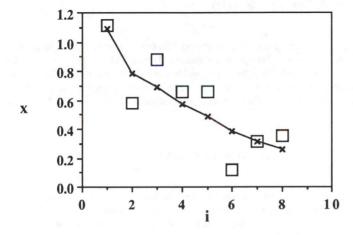

Fig. 5.25 - Smoothing of data in Table 5.9 using a Kalman filter.

More elaborate Kalman filter equations can be derived for multi-input systems where several measurements are made in order to estimate one parameter: these measurements may, for example, be several wavelengths which are used to measure the concentration of a compound by uv / visible spectroscopy. The references in the bibliography expand on the use of Kalman filters in more complex situations.

REFERENCES

Section 5.2
R.G.Brereton, *Chemometrics Int. Lab. Systems*, **1**, 17 (1986)
A.G.Marshall (Editor), *Fourier, Hadamard and Hilbert Transforms in Chemistry*, Plenum, New York, 1982
P.R.Griffiths (Editor), *Transform Techniques in Chemistry*, Heyden, London, 1978

R.W.Remirez, *The FFT. Fundamentals and Concepts*, Prentice Hall, Englewood Cliffs, NJ, 1985
A.G.Marshall and F.R.Verdun, *Fourier Transforms in NMR, Optical and Mass Spectrometry: A User's Handbook*, Elsevier, Amsterdam, 1989

Section 5.3
A.Savitsky and M.J.E.Golay, *Anal. Chem.*, **36**, 1927 (1964)
W.E.Blass and G.W.Halsey, *Deconvolution of Absorption Spectra*, Academic Press, New York, 1981
P.A.Jansson (Editor), *Deconvolution with Applications to Spectroscopy*, Academic Press, Orlando, 1984
S.Goldman, *Information Theory*, Dover, New York, 1953
P.H.Van Cittert, *Z. Phys.*, **69**, 298 (1931)

Section 5.4
J.Skilling, S.Sibisi, R.G.Brereton, E.D.Laue and J.Staunton, *Nature*, **311**, 446 (1984)
J.Skilling (Editor), *Maximum Entropy and Bayesian Methods*, Kluwer, Dordrecht, 1989
E.T.Jaynes, *Papers on Probability, Statistics and Statistical Physics*, Reidel, Dordrecht, 1982
J.P.Burg, *Maximum Entropy in Spectral Analysis*, PhD Thesis, Stanford, CA, 1975
P.A.Jansson (Editor), *Deconvolution with Applications to Spectroscopy*, Academic Press, Orlando, 1984

Section 5.5
A.Savitsky and M.J.E.Golay, *Anal. Chem.*, **36**, 1927 (1964)

Section 5.6
C.Chatfield, *The Analysis of Time Series: An Introduction*, Chapman and Hall, London, 1984
J.C.Lindon and A.G.Ferrige, *Prog. NMR Spectrosc.*, **14**, 27 (1980)
J.K.Kauppinen in G.E.Vanasse (Editor), *Spectrometric Techniques, Volume III*, Academic Press, New York, p. 199 (1983)
H.C.Smit in B.R.Kowalski (Editor), *Chemometrics: Mathematics and Statistics in Chemistry*, Reidel, Dordrecht, p. 225 (1984)
H.C.Smit, *Chemometrics Int. Lab. Systems*, **8**, 15 (1990)
H.C.Smit, *Chemometrics Int. Lab. Systems*, **8**, 29 (1990)

Section 5.7
S.C.Rutan, *Chemometrics Int. Lab. Systems*, **6**, 191 (1989)
S.D.Brown, *Anal. Chim. Acta*, **181**, 1 (1986)
S.C.Rutan, *J. Chemometrics*, **1**, 7 (1987)

6

Multivariate signal processing

6.1 MULTIVARIATE INSTRUMENTAL SIGNALS

Over the last few years the capabilities of laboratory instrumentation have increased substantially. In the 1970s substantial time and effort was required to obtain single univariate spectroscopic or chromatographic traces. Faster instruments, new methods of data acquisition (e.g. Fourier spectrometers) and the availability of cheap on-line computing now permits the experimenter much greater freedom to acquire data. This means that *sets* of measurements are readily available. These sets of measurements may be of two types as follows.

(1) A series of measurements on a related set of samples. For example, each sample may consist of a mixture of compounds in differing ratios. The objective of the analysis might be to determine the amount of each compound in each sample. We know that there are common features in each sample, namely information arising from each compound. This problem is one of mixture analysis.

(2) Hyphenated analytical techniques. These approaches, often involving joint chromatographic and spectroscopic analysis (e.g. LC-MS, Diode array HPLC, GC-MS), are increasingly common. A two dimensional dataset of spectra and chromatograms is obtained. These measurements are of a set of chromatograms at various spectroscopic wavelengths or frequencies. So if each chromatogram is monitored at 64 wavelengths, the datasets consist of 64 chromatograms. This problem is analogous to that of mixture analysis: if we were to analyse 64 samples and take the chromatogram of each sample we would, likewise, produce a similar sized dataset. Hence related techniques are employed both for mixture analysis and processing information acquired by hyphenated instrumentation.

In these cases we have available *multivariate* instrumental data. It is possible, of course, to treat each chromatogram or spectrum independently, and use methods such as those discussed in Chapter 5 to improve the quality of the data. However, these approaches omit certain information, particularly that of trends common to each univariate trace.

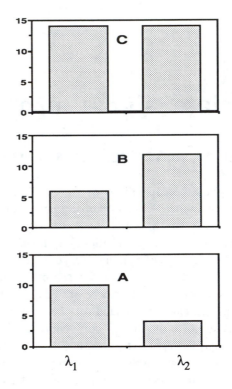

Fig. 6.1 - Spectral intensities measured at two wavelengths for compounds A, B and C (Table 6.1).

Consider a simple problem in which a mixture of three compounds is observed at two wavelengths, λ_1 and λ_2. The proportions of compound p, given by α_p add up to 1, i.e.

$$\sum_{p=1}^{p=3} \alpha_p = 1 \tag{6.1}$$

The pure spectra for each compound is tabulated in Table 6.1 and illustrated in Fig. 6.1.

Table 6.1 - Intensity of spectra of three compounds at two wavelengths

	Compound		
	A	B	C
λ_1	10	6	14
λ_2	4	12	14

If a mixture consists purely of compound A, there is only one possible value of λ_1 and λ_2 which could be represented by a point (10, 4) in two dimensional space where each axis represents one wavelength. For mixtures consisting only of compounds A and B, λ_1 and λ_2 will lie on a straight line as illustrated in Fig. 6.2. The distance along this line will determine the proportion of compound in the mixture. For mixtures consisting of three components, the possible values of λ_1 and λ_2 will lie in a triangle (Fig. 6.3). The distance of a spectrum from each vertex of the triangle is related to the proportion of components in the mixture.

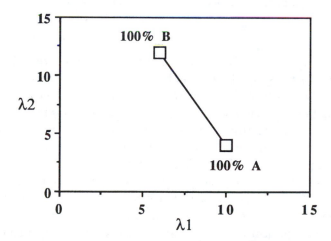

Fig. 6.2 - Graphical representation of spectra consisting of a mixture of compounds A and B.

It is impossible to uniquely determine the proportion of four compounds in a mixture by observing only two wavelengths: $p-1$ wavelengths are required for the determination of the proportion of compounds in a p-component mixture, providing the proportions of components are normalized, i.e. add up to 1.

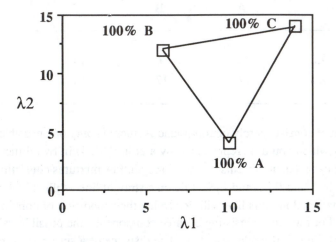

Fig. 6.3 - Graphical representation of spectra monitored at two wavelengths consisting of mixtures of compounds A, B and C. Each spectrum can be represented by a point in the triangle.

We have already shown how this information may be visualized. The number of dimensions equals the number of wavelengths observed or the number of *sensors*. All observations fall within an $p-1$ dimensional cross-section within this space, providing the spectra of each compound are linearly additive.[1] The distance from the spectrum (represented as a point in m dimensional space, where m is the number of sensors) of each pure component is related to the proportion of each constituent in the mixture.

The methods for calculating the proportion of a compound in a mixture are very similar to PCA discussed in Section 4.2.[2] One important reason for using PCA is that measurements are never free of noise: so a three component mixture might not exactly project onto a triangle and it is important to understand methods for determining the number of significant components. Most of this chapter will be devoted to discussing enhancements to principal component type approaches for quantifying components in a mixture. A typical experiment might involve observing a four component mixture at 100 wavelengths and so provide substantially more information than in the case discussed above. The data

[1] In this chapter we assume that there are no "interactions" unless specified.
[2] We assume that the reader will refer back to Section 4.2 if necessary.

analytical strategies depend on what is required from the experiment and what is known about the components beforehand. Are the spectra of the pure components known? Are we interested in the relative or the absolute amounts of each component? Are we interested in determining the spectra of each component from a series of mixtures? How many components are actually in a mixture?

6.2 FACTOR ANALYSIS: DETERMINING THE NUMBER OF SIGNIFICANT FACTORS

Factor analysis is one of the most used techniques in chemometrics. The original ideas derive from several different schools of thought. Perhaps the earliest use of factor analysis was by psychometricians in the 1940s. Verbal and written tests were developed to estimate factors such as intelligence, ability and aptitude. These tests provided matrices with the rows corresponding to people and the columns to scores or answers to the tests. Statistical approaches were then used to condense this information to various principal factors. Most psychometric methods were originally based on PCA.

   As an illustrative chemical example, consider a system where we measure two spectroscopic peaks arising from a single compound. The theoretical spectrum consists of peak x_1 of intensity 1 and peak x_2 of intensity 2 (Fig. 6.4). A simple experiment might involve recording the intensities of x_1 and x_2 at different concentrations of the compound: as usual noise is imposed upon the signal. Table 6.2 is a simulation of the experimental data: these data now no longer exactly lie on a straight line and it is now no longer certain that there is only one compound in the "mixture". From these data we may wish to answer three questions.

(1) How many compounds are there in the mixture?
(2) What is / are the pure spectra of these compounds?
(3) What is the concentration of each compound in each experiment?

   In order to answer the questions above, it is usual to try to reduce the dimensionality of the data via PCA. The first step (Section 4.2) is to find x_1' and x_2' such that

$$x_1' \quad = \quad \cos\theta \, x_1 + \sin\theta \, x_2 \qquad\qquad (6.2)$$
$$x_2' \quad = \quad -\sin\theta \, x_1 + \cos\theta \, x_2 \qquad\qquad (6.3)$$

where $\sum_{i=1}^{i=N} x_2'^2$ is minimized. This is equivalent to finding the value of b_1 for the best fit straight line[1]

$$\hat{x}_2 \quad = \quad b_1 \hat{x}_1 \qquad = \qquad \tan\theta \, \hat{x}_1 \qquad\qquad (6.4)$$

---

[1] The coefficients $\cos\theta$ and $\sin\theta$ are often called loadings and the values of x_i' are called scores as discussed in greater detail below.

Table 6.2 - Two peaks in the presence of noise

Expt.	x_1	x_2
1	0.8404	2.0296
2	1.8982	4.1676
3	3.1831	5.8166
4	4.1665	7.9562
5	5.3620	9.8650
6	5.9200	11.6624
7	7.0987	13.8794
8	8.1588	16.3139
9	8.9828	18.0685
10	10.3643	20.3791

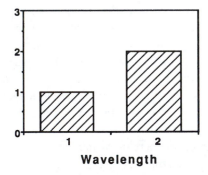

Wavelength

Fig. 6.4 - Theoretical spectrum of a compound measured at two wavelengths.

For the data of Table 6.2, we find that $b_1 = 1.9736$ and so $\theta = 1.1018$ radians or 63.13°. It is easy to compute the two principal components,[1] via Eqs (6.2) and (6.3); the best fit straight line is given by

$$\hat{x}_1 = \cos\theta\, x_1' \qquad\qquad (6.5)$$
$$\hat{x}_2 = \sin\theta\, x_1' \qquad\qquad (6.6)$$

---

[1] See Sections 4.2 and the appendix for a discussion of how to calculate principal components.

since the first principal component corresponds to the distance along the best fit straight line.

The principal components and best fit straight line are tabulated in Table 6.3. We can compute various sums of squares: the sums of squares of the two principal components are identical to the sums of squares of the raw data (1935.18). However the sums of the squares of the estimated (best fit) values of $\hat{x}_1$ and $\hat{x}_2$ add up to the sum of squares of the first principal component only. This sum is fairly close to the overall sum of squares, however, which is expected if there is only one compound present. The second factor represents noise.

Table 6.3 - Principal components and best fit straight line (using principal components criteria) for data of Table 6.2

Expt	x_1'	x_2'	$\hat{x}_1$	$\hat{x}_2$
1	2.190	0.167	0.990	1.954
2	4.576	0.190	2.068	4.082
3	6.627	−0.210	2.995	5.912
4	8.980	−0.121	4.059	8.011
5	11.223	−0.324	5.073	10.012
6	13.079	−0.010	5.911	11.667
7	15.589	−0.059	7.046	13.906
8	18.240	0.096	8.244	16.271
9	20.178	0.154	9.120	17.999
10	22.863	−0.034	10.334	20.395

It is fairly obvious that a single factor model is adequate since the ratio between the sums of squares of the first principal component (often called the *eigenvalue*) to the second is very large. This will not always be the case and a test for the number of significant factors will depend on the ratio of these successive sums of squares. At this point we need to define our terminology in greater detail. As is common in chemometrics there are no universal conventions so the reader much check carefully the terminology used by different authors. The conventions we employ are that a given dataset consists of a number of observations or rows ($=r$) and a number of variables or columns ($=c$). In the example of Table 6.2, $c=2$ and $r=10$. Although most chemometrics software has been developed using matrix notation, which is convenient for programming purposes, it is not, of course, necessary to understand the mechanics of the matrix algebra to be able to interpret the output from chemometrics programs, but some appreciation of the use of such notation is useful for understanding chemometrics papers and texts. The raw data for factor analysis is often referred to as an $r \times c$ matrix: the data of Table 6.2 forms a 10×2 matrix. The nth value of principal component j (x'_{jn}) is often called the *score* of object n for the given principal component. For example, the score of the fourth experiment (object) for principal component 1 is 8.683 in our example. The *loadings* are merely the coefficients in Eqs 6.2 and 6.3 for a two principal component model. In a more general case the loading of the jth principal component is the coefficient α_{ji} in the equation

$$x'_j \quad = \quad \sum_{i=1}^{i=c} \alpha_{ji} x_i \qquad\qquad (6.7)$$

Therefore each variable and object is associated with a loading and a score for each principal component. The sum of squares for each principal component is sometimes called an eigenvalue, which we denote as κ_j (so that, in the example above $\kappa_1 = 1952.91$).

Table 6.4 - Sums of squares for variables in Tables 6.2 and 6.3

Data	Parameter				Total
Raw data	x_1	400.67	x_2	1552.51	1953.18
PCs	x'_1	1952.91	x'_2	0.27	1953.18
Estimated values	$\hat{x}_1$	400.89	$\hat{x}_2$	1552.02	1952.91

Experimental measurements, though, are associated with noise as well as pure signal. If c variables are measured a major objective of factor analysis is to determine the p significant factors required to adequately model the data: ideally p is equal to the number of components in a mixture. The remaining factors should represent pure noise. It is usual to define the sum of squares of the first p significant factors

$$S_{\text{model}} \quad = \quad \sum_{j=1}^{j=p} \kappa_j \qquad\qquad (6.8)$$

and the sum of squares of the remaining factors

$$S_{\text{error}} = \sum_{j=p+1}^{j=c} \kappa_j \qquad (6.9)$$

Of course there are other ways of decomposing the sum of squares but ideally the ratio $S_{\text{model}} / S_{\text{error}}$ should be large if p is chosen correctly. In our example there are only two possible sums, but in a more realistic situation we might measure 20 spectroscopic intensities and try to reduce these data to three or four principal components representing three or four compounds. The chemometrician is interested in determining the optimum value of p that best models the data.

There are several ways of using the sums of squares in Eqs (6.8) and (6.9) to assess the significant number of factors. Much of the original theory of errors in chemical factor analysis is due to Malinowski. He uses the concept of "pure data" which is the error free underlying data, which we will denote by ξ_{in}. If the model is perfect the "pure data" should equal the predicted data, but there is normally an error between pure and estimated data. The errors are, as usual, additive, so that

$$x_{in} = \hat{x}_{in} + \eta_{in} = \xi_{in} + \varepsilon_{in} \qquad (6.10)$$

We never know the values of ξ_{in} so never know the true error between the observed and estimated data. However, chemometricians should attempt to make use of simple, intuitive assumptions. Most of the methods for data analysis described in this text will not function well if the assumptions are far from the truth. A possible assumption is that the two error terms (η_{in} and ε_{in}) in Eq. (6.10) are roughly equal in size. We can define three errors from differences between terms in Eq. (6.10) (Table 6.5). The sum totals of these errors are often referred to by XE, RE and IE: we can only observe the extracted error. The other two errors are indeed properties of the system but not experimentally measurable.

Table 6.5 - Main factor analysis error terms

Error	Symbol	Terms	Common terminology
$x_{in} - \hat{x}_{in}$	η_{in}	p	Extracted Error (XE)
$x_{in} - \xi_{in}$	ε_{in}	c	Real Error (RE)
$\hat{x}_{in} - \xi_{in}$	$\varepsilon_{in} - \eta_{in}$	$(c-p)$	Imbedded Error (IE)

Note : It is assumed that there are c variables measured and p significant principal components.

Various approaches to estimating the number of significant factors, and so the number of relevant components in a mixture, depend on comparing these error terms. Malinowski's theory of errors assume that the terms are roughly of the same order of magnitude and that

$$RE^2 = XE^2 + IE^2 \qquad (6.11)$$

It can be shown that, under such circumstances,

$$RE = \sqrt{c/(c-p)}\ XE \qquad (6.12)$$

and

$$IE = \sqrt{p/(c-p)}\ XE \qquad (6.13)$$

It is, therefore, possible to estimate the various error terms from the observed extracted error for different numbers of principal components or factors. If we are testing whether p principal components adequately model a dataset, the extracted error is merely the sum of squares of principal components $p+1$ to c. In the example discussed above we expect that one principal component alone adequately models the data, so $p=1$ whereas $c=2$. The value of XE is merely $\sqrt{0.27}\ /rc = 0.115$; hence the *estimates* of IE = 0.115 and RE=0.163.

We can never be certain of the nature of the "pure" (noise free) data, merely hope that the model is a good approximation. In fact the data of Table 6.2 was obtained as follows.

(1) A pure dataset of the form $\xi_{1n} = n$, $\xi_{2n} = 2n$, where ξ_{in} is the nth row (observation) of the parameter ξ_i, was used.

(2) A Gaussian noise distribution of mean = 0 and standard deviation = 0.2 was imposed on these data to give the simulated dataset.

So the value of ξ_1 for the pure data used to simulate data experiment 7 is 7: the value of ξ_2 is 14 for the corresponding experiment. Obviously the experimenter has no way of knowing this in advance, but he or she hopes that the estimated values are in reasonable agreement with the true values. It is possible to compute the true values of IE and RE using this information, and these are 0.173 and 0.207 respectively. Although these true values do not exactly equal the estimated values of these errors, they are likely to show similar trends, and in the case discussed above are similar in magnitude. The reason for the disagreement is that errors are not necessarily evenly distributed. There is a large literature on the use and estimation of various error indicator functions in factor analysis but a key to the effectiveness or otherwise of different approaches lies in the nature of the noise. Often the experimenter does not study underlying noise distributions so the use of these approaches is essentially empirical. Under some circumstances one method might be effective (e.g. for a particular kind of spectroscopy), so, historically, that method is employed for the particular application without querying its theoretical basis. Although this may seem poor science, an experimentalist must evaluate the time and effort required to contrast different methods and to study the nature of the underlying noise. There can be no general guidelines.

Normally the magnitude of various error terms is calculated according the number of principal components (p) in the model. There are several criteria for estimating the value of p that represents the true number of components. Three of the most popular are as follows.

(1) *Imbedded error*. The value of IE decreases until the true number of components has been found and then increases again. Normally IE must be estimated from XE as described above.

(2) *Malinowski indicator function* (IND). This function also is supposed to reach a minimum when the true number of components has been found. It is defined by

$$IND = RE/(c-p)^2 = XE\sqrt{p/(c-p)}/(c-p)^2 \qquad (6.14)$$

provided Eq. (6.13) holds approximately. This function is found to give sharper minima than the imbedded error in many experimental situations.

(3) *Malinowski F-test*. This approach is somewhat easier to understand statistically: the reader is referred to Section 2.5 for a more detailed discussion of F-tests. To determine the number of significant factors, the ratio

$$F_{1,c-p} = (c-p)\,\kappa_p / \sum_{i=p+1}^{i=c}\kappa_i \qquad (6.15)$$

is calculated. The significance of factor p relative to the remaining $c-p-1$ factors can then be estimated by using a table of F-ratios. Recently several elaborations to this approach have been reported. This method gives a probability that principal component p is a genuine component rather than merely arising from noise.

Most of the approaches above are most effective in areas such as spectroscopy where noise is often normally distributed. An alternative approach is called *cross-validation*. The first chemical applications were reported by S.Wold and coworkers, although it is well known in statistics. There are several implementations of this method, but the usual approach is to see how well a principal component can predict missing values. If, for example, there are 100 observations, can a principal component model based on 90 of these observations predict the remaining 10 observations? Cross-validation to test the significance of the pth principal component works as follows.

(1) Remove $\alpha\%$ (typically about 10%) of the points in a dataset. Calculate the pth principal component using $(100-\alpha)\%$ of the data.

(2) Add this principal component to the model of the first $p-1$ components which is used to *predict* the missing $\alpha\%$ of the dataset, and calculate the total sum of square error between the predicted and experimental data.

(3) Remove a *different* $\alpha\%$ of the data and repeat steps 1 and 2. Continue until each datapoint has been removed once, so, for example, if $\alpha=10\%$, steps 1 and 2 are repeated 10 times.

(4) The total error sum of squares calculated in step 2 is computed. This is often referred to as the *predicted residual error sum of squares* for the pth principal component (PRESS(p)).

(5) If $p > 1$, where p is the current principal component, compare PRESS($p-1$) to PRESS(p). The comparison of these values is used to indicate whether sufficient components have been calculated to adequately model the data. A simple criterion is that p components are adequate to model the data if PRESS(p) > PRESS($p-1$). A different approach used is to compare the value of the predicted residual error sum of squares to the actual residual sum of squares [RSS($p-1$)] using the previous component.

(6) If insufficient components have been calculated continue with step 1 for the next component.

There are several ways of deleting $\alpha\%$ of the data. A favoured approach in chemometrics is to delete values on the diagonals of an $r \times c$ matrix (Fig. 6.5), but random number schemes may also be employed.

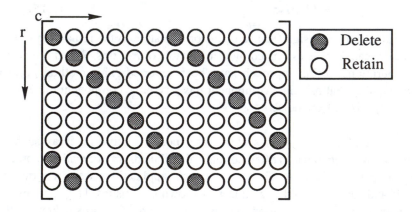

Fig. 6.5 - Illustration of cross-validation.

One advantage of cross-validation is that it allows principal components to be calculated one at a time, and the calculation stops once an adequate model is established. This was of advantage in early microprocessors: a typical dataset might consist of 20 variables, but only three or four principal components are significant. The Malinowski type approaches of estimating and comparing residual errors require calculation of all the principal components, which was more suited to off-line mainframe computers. With the advent of faster on-line microcomputers this is no longer such a serious limitation. A second advantage of cross-validation is that it selects components with greater *interpretative* value. Sometimes the first principal component might be strongly influenced by size; for example, we may be measuring the concentration of chemicals in a series of samples, and the most important factor influencing the spectra is the amount of sample used. This may dominate the analysis. However smaller but more interesting differences between samples may be buried in the second and later principal components, although the size of these components (i.e. the sum of squares for each component) may be very small relative to the first component.

Generally cross-validation appears to be less sensitive to difficulties caused by skew and non-normal noise distributions. It is probably the preferred method in applications such as geochemistry, pharmaceutical chemistry, environmental chemistry and so on where normal noise distributions are less likely. There are, however, no clear guidelines as to which approach is preferable. Because different groups have worked on different types of data, there are often conflicting claims in the literature.

6.3 FACTOR ANALYSIS: TARGET TESTING AND ROTATIONS

Within the context of chemical mixture analysis, factor analysis can be used to determine whether an individual compound is present in a mixture, and if so, in what quantities.

To illustrate this, let us consider a simple example, where a mixture of two compounds is monitored at three wavelengths. The raw spectral intensities of each of these compounds are given in Table 6.6.

Table 6.6 - Spectra of two compounds monitored at three wavelengths

Compound	x_1	x_2	x_3
A	1	2	3
B	4	1	2

Note : x_i is the intensity of one "unit" of the appropriate compound at wavelength i.

The observed intensities at each wavelength are the sum of intensities due to each of the individual compounds. So, for example, a spectrum of the form $\mathbf{x} =$

(6, 5, 8),[1] arises from 2 "units" of Compound A and 1 "unit" of Compound B. It is clear that there are only certain possible spectra consistent with this two component mixture. The spectrum $x = (3, 3, 3)$ cannot possibly arise from this mixture.

It is usual to represent this information geometrically. A spectrum can be considered as a point in three dimensional space with the axes x_1, x_2 and x_3. However, all spectra arising from these compounds must lie on a plane because there are only 2 degrees of freedom or two independent axes. Ideally the position in this plane will tell us the relative amounts of Compounds A and B. It is fairly easy to define the equation for this plane,[2] which is given by

$$x_1 + 10x_2 - 7x_3 = 0 \tag{6.16}$$

For example, the co-ordinates (6, 5, 8) satisfy this condition, whereas the co-ordinates (3, 3, 3) do not.

Table 6.7 - Possible axes for the plane defined by the spectra in Table 6.6

	Unscaled x_1'	x_2'	Scaled x_1''	x_2''
x_1	10	1	0.9950	0.0569
x_2	-1	10	-0.0995	0.5687
x_3	0	(101/7)	0.0	0.8206

Chemometricians normally need to calculate the orthogonal axes of a plane rather than merely an equation of the form of Eq. (6.16), as discussed in Section 4.2. There are an infinite number of pairs of axes, and in the absence of experimental datapoints we cannot choose between these. One solution is to determine an axis in this plane that contains $x_3 = 0$ and then another axis in the plane orthogonal to this, using Eq. (4.12). It is usual, but not essential, to scale the axes so that $\sqrt{(x_1^2 + x_2^2 + x_3^2)} = 1$. These axes are given in Table 6.7. The new axes can be represented by vectors in the original 3 dimensional space, and define the new two dimensional "plane". This projection onto a plane is

[1] We employ simple vector notation so that x stands for (x_1, x_2, x_3).
[2] The equation for the plane can be calculated by simple geometry. The plane must contain the three points (0, 0, 0), (1, 2, 3) and (4, 1, 2). If the equation is of the form $Ax_1 + Bx_2 + Cx_3 = 0$, then $A + 2B + 3C = 0$ [compound A] and $4A + B + 2C = 0$ [compound B].

illustrated in Fig. 6.6. All the experimental measurements will fall within this plane providing there is no noise and the spectra consist only of the two components given above. Since the amount of each compound must be positive, spectra will only occupy a region of this plane. We shall return to this latter limitation below.

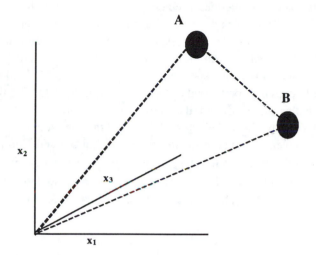

Fig. 6.6 - Spectra of compounds A and B (Table 6.6) in three dimensional space, each axis representing one variable.

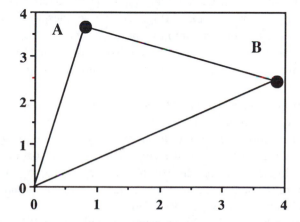

Fig. 6.7 - Projection of data of Table 6.6 onto a plane, with the scaled axes as defined in Table 6.7.

In a typical series of experiments in which we measure three wavelengths in differing relative concentrations of the two compounds, it is possible to find a plane and so reduce the raw data to two principal components, but the individual spectra of the the two pure compounds might be unknown. In more complicated situations such a naturally occurring mixture might consist of four or five compounds (so the spectroscopic measurements are projected onto four or five dimensions) and 20 or 30 wavelengths might be monitored. If the identity of the pure compounds is unknown can we test various possible spectra and see whether these are consistent with an individual component in the data? In the case above we may wish to test the spectra $x = (1, 2, 3)$ and $x = (4, 1, 2)$ to see whether these could possibly correspond to a component of the data. The way to do this is to see whether these spectra fall exactly onto the plane.

In the plane defined by the scaled axes of Table 6.7 the spectrum of one unit of Compound A is given by the co-ordinates $x'' = (0.80, 3.65)$ and Compound B by $x'' = (3.88, 2.44)$. This is quite easy to demonstrate since $0.995 \times 0.80 + 0.0569 \times 3.65 = 1$ (the intensity of 1 "unit" of Compound A) and so on. The calculation for Compound A can be expressed in matrix notation as follows:

$$
\begin{pmatrix}
0.995 & 0.0569 \\
-0.0995 & 0.5687 \\
0.0 & 0.8206
\end{pmatrix}
\cdot
\begin{pmatrix}
x''_1 \\
x''_2
\end{pmatrix}
=
\begin{pmatrix}
1 \\
2 \\
3
\end{pmatrix}
\tag{6.17}
$$

and a similar matrix equation can be given for Compound B.

The new co-ordinates for Compounds A and B can be represented by two lines in the principal component plane (Fig. 6.7). These lines are the *projection* of the spectra for each compound onto this plane. In Section 4.2 we discussed projections in the context of dimensionality reduction. Similarly the dimensionality of the mixture space is equal to the number of components in the mixture.

The two axes of the plane have no physical significance but the two projected spectra have a physical meaning. In fact the principal co-ordinates will depend, in part, on the distribution of the samples rather than on the pure spectra, but should, in the absence of noise, be in the same plane as the vectors that represent the spectra. It is possible to *rotate* the axes so that they overlap exactly with the vectors of the pure spectra. This is the basis of *target transform factor analysis* (TTFA), sometimes called *target factor analysis*.

In reality the problem is slightly more complicated. Generally far more than three wavelengths are observed and there are usually more than two components in a mixture, so factor analysis becomes a problem of projecting c measurements onto p dimensions. The first step is to find the number of significant factors using approaches discussed in Section 6.2. Once this is determined, the first p principal components can be calculated. Normally, as discussed in Section 4.2, these components will be unique, and dependent on the experimental observations. The next step might be to test possible compounds or spectra that could be present in the mixture. This is performed by rotating the spectrum of a compound in the p-dimensional factor space until it is overlaps as closely as possible with one of the axes. The *residual sum of square* error between the rotated test factor and the principal component axes is used as a measure of goodness of fit. The smaller this error, the more likely this spectrum is to be consistent with one component of the mixture. This method is also an example of *principal component regression* as a test factor (the spectra of a candidate compound) is regressed (using rotation) against the principal component: this method will be discussed in more detail, in another context, in Section 6.4.

There are many variances on TTFA, including ITTFA (*iterative target transform factor analysis*), in which the factors are determined iteratively. The advantage, to the chemist, of these approaches are that the factors are given a physical meaning and these methods have been very effective in pulling out components from mixtures. Another aim of factor analysis is to quantify how much of a given compound is in a mixture. This is not always a simple problem as we do not always know whether the mixtures encompass pure compounds or not. In Section 6.1 we showed that λ_1 and λ_2 will lie on a straight line in the case of a mixture of two compounds whose proportions add up to 1.[1]

In the example discussed above *all* experimental measurements should lie between the straight lines (vectors) in two dimensional factor space corresponding to compounds A and B. Measurements outside this region correspond to negative concentrations of compounds and so are physically meaningless. The position of a spectrum (represented by a point in this plane) relates directly to the relative proportion of the components in a mixture.

There are several difficulties with this apparently simple approach to estimating the amount of a component in a mixture. The most important one is that there is normally no certain information as to the nature of the spectra of each of the individual components. A naturally occurring series of mixtures will not necessarily efficiently span the mixture space. For example, the concentration of a compound might vary between 30% and 80% in a series of mixtures – so no mixture contains one single pure compound. In pharmacological, geochemical, environmental etc. problems, this situation will be quite common. In Fig. 6.8 we illustrate the same series of mixtures as in Fig.

---

[1] In the example discussed immediately above we considered mixtures containing any possible proportions of compounds: some experiments are used to measure relative and others with absolute quantities – it is always important to establish which type of experiment is most useful in a particular situation. Scaling is discussed in further detail in Section 7.2.

6.7, but now the upper and lower limits of compound A are 80% and 30% respectively. The best and most realistic approach to determining the true limits to the mixture space is by using an approach such as TTFA as discussed above. The projection of the target spectrum will still lie in the principal component plane, so it is possible to test various possible compounds. In this way the chemometrician uses knowledge he or she has about the system such as the possible identity and spectra of individual components.

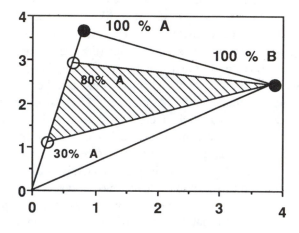

Fig. 6.8 - Allowed region in the space defined in Fig. 6.7 if compound A varies between 30% and 80%.

Sometimes mixtures are compositional, that is the proportions of each component add up to 1. A typical example might be the proportion of solvents in a mixed solvent system. Exactly the same methods for factor analysis may be used to determine the spectra of individual components, but the method for calculating the proportion of each component in a mixture is slightly different. As discussed in Section 6.1, the spectra of the mixtures will be contained within a $p-1$ dimensional simplex[1] and the distance from the vertices of the simplex will be determined by the relative proportions of the components in the mixture. Of course, the axes do not have to correspond to single wavelengths but might correspond to principal components if more than $p-1$ wavelengths are recorded (which is normally the case). Naturally, it is important to know whether the vertices of the experimentally determined simplex correspond to pure compounds or to a mixture. In Section 4.6, we discussed a mixture triangle in which the vertices were not pure components.

---

[1]See Section 2.3 for the definition of a simplex.

Putting knowledge to use, such as the possible identities and spectra of components in a mixture, is an important aspect of chemometrics and helps in quantitating mixtures.

The methods described above are used to identify factors by target rotations. There is a long history of factor analysis and many of the earliest applications in psychometrics and geology also involved rotations, but different from chemical applications in that individual factors rarely have a precise, physically interpretable, meaning. They were used as ways of simplifying the data, probably giving a better picture than principal components. These methods are forms of *abstract factor analysis* and involve rotation subject to various criteria. In the discussion above we plotted graphs of the principal component scores. It is also possible to plot the loadings: in the example above the loading of x_1 is (0.995, 0.057) (see Table 6.7). Normally, of course, the principal components are determined from a real set of mixtures, and, as explained in Section 4.2, there is a unique set of principal component axes with a unique origin. In the examples above we were only concerned with *projections* in the absence of data. Ideally, unless noise significantly distorts the data, the same principal component space is determined whatever experiments are performed, providing the constituents and spectra are the same in each set of spectra, differing only in the origin and axes. In the discussion below we assume that the rotations take place in loadings space.

Probably the best known approach (available in most commercial factor analysis software) is called *varimax* rotation. The principal components are rotated so that the total sum of squares of the loadings along each new axis is maximized. This rotation is an example of an *orthogonal rotation*: the geometry of the factors remains the same – so that if two axes are initially at right angles to each other they remain at right angles after rotation. It is unlikely that two physically significant factors are at right angles to each other so varimax rotation is most unlikely to extract spectra of individual compounds, but could be useful, for example in determining sources of pollution. A separate class of rotations are *oblique rotations* where the relative geometry of the axes is not preserved. Above we discussed methods for determining the proportion of compounds in a two component mixture. Since the axes corresponding to each component in the mixture were not at right angles to each other, the principal components are unlikely to correspond exactly to individual components. Two principal components, initially at right angles to each other, might be rotated to yield two factors that are no longer at right angles to each other. There are several different criteria for these rotations and a large number of papers have been written contrasting the merits of these approaches. The interested reader is referred to the bibliography at the end of this chapter for further details.

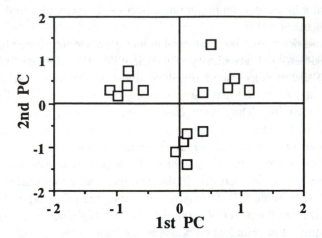

Fig. 6.9 - Loadings plot for Table 6.8.

Table 6.8 - Principal component loadings for 15 peaks.

Peak (variable)	PC1	PC2
x_1	0.8927	0.5517
x_2	0.1139	−1.3961
x_3	0.1208	−0.6874
x_4	1.1244	0.2916
x_5	−0.9755	0.1747
x_6	−0.5727	0.2936
x_7	0.7774	0.3537
x_8	0.3790	−0.6445
x_9	0.0553	−0.8869
x_{10}	−0.8177	0.7452
x_{11}	0.3809	0.2526
x_{12}	−1.0990	0.3041
x_{13}	−0.0610	−1.1047
x_{14}	0.5095	1.3366
x_{15}	−0.8278	0.4158

Finally, one other approach to factor analysis, developed by Windig, largely for interpretation of pyrolysis mass spectra, will be described. This involves calculating a *variance diagram*. Consider a series of spectra consisting of 15 peaks. It is suspected that the spectra consist of three components and that the data is compositional. Therefore the data should reduce to two principal components with the three pure constituents at the corners of a triangle in the reduced principal component space. Loadings for these 15 peaks are tabulated (Table 6.8).

The graph of these loadings is illustrated in Fig. 6.9. From this figure, it appears that the loadings are "clustered" into various groups. In a technique such as pyrolysis mass spectrometry we might expect this type of pattern for the following reasons. First, the each compound normally gives rise to certain characteristic masses. Second, peaks are well defined quantities corresponding to different masses. This technique would not work well in uv / visible spectroscopy for example: spectra would normally overlap completely and not give rise to characteristic wavelengths.

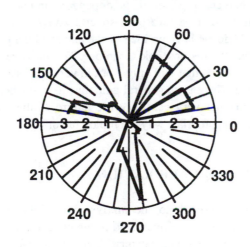

Fig. 6.10 - Variance diagram corresponding to the loadings plot of Fig. 6.9, using 20° divisions, as discussed in the text.

A variance diagram can be constructed as follows.

(1) Divide the loadings plot into angular sections of about 10° or 20°. In the example below we use 20°.

(2) Take all the points whose loadings fall within each section, and sum the squares of the loadings.

(3) Replot the sums of squares of loadings in each section to give a variance diagram.

(4) The directions in which the loadings are maximum should correspond to pure constituents.

A variance diagram for the loadings plot in Fig. 6.9, is illustrated in Fig. 6.10. It suggests that there are three components at roughly 50° (because of the digitization and small number of variables this component appears split into two subcomponents), 170° and 270°. These directions can then be used to locate the pure components and also derive the spectra for these components. One advantage of this simple graphical method is that there is a check on the overall process of principal component reduction. A three component compositional mixture should be represented by two principal components, so we expect three maxima in this plane. If there are a different number of maxima, this suggests a problem in the analysis: either we have determined the wrong number of components or there is too much noise in the measurement technique.

The discussion of factor analysis techniques given above is not exhaustive. There are a huge number of possibilities, dependent, in part, on what is known about the system in advance (are pure spectra known? can pure spectra be guessed and tested? do we have no idea of the nature of the pure spectra?) and, in part, on the particular spectroscopic technique. The variance diagram approach is an example of a method that will only work in very specialized circumstances, but is very effective. There is no general guideline as to the best approach and it is important to consider very carefully the nature of the experiment before deciding on the most appropriate way to employ factor analysis.

6.4 MULTIVARIATE CALIBRATION

Another important approach to the analysis of mixtures is *multivariate calibration*. Quite often we are interested in the concentration of a single, or a small number, of compounds in a mixture. A good example is the estimation of chlorophyll pigments in natural water samples that contain algae: the amount of chlorophyll *a* yields information about the productivity of the organisms in the water sample, and so is related to light intensity, time of year, pollutants and so on. It is relatively easy to extract the pigments and take their uv / visible spectra. It is also easy to take the spectra of pure chlorophylls, but the natural pigment extracts are bound to be contaminated by other compounds. One approach to estimating the concentration of chlorophylls would be to add known amounts of chlorophyll *a* and monitor the change in the spectra. These can be used to *calibrate* the chlorophyll concentration to the spectra and so relate two types of variables, namely spectroscopic intensities and concentrations.

The simplest form of calibration is called *univariate calibration*. In such a case a single wavelength or detector is monitored and the change in intensity as the concentration of a compound is varied, is used to construct a linear model of intensity *versus* concentration. The better the straight line fit, the more useful the method. Unfortunately in most practical situations, there are interferants, that is compounds and solvents that also absorb in the regions where the compound of interest absorbs. Because the concentration of these interferants is not always the same in each sample, it is not too easy to correct for them using univariate calibration. In Table 6.9, we present some data where three peaks are monitored (x_1, x_2 and x_3) as different concentrations of a compound (y) are added to the samples.

Table 6.9 - Calibration example

y	x_1	x_2	x_3
0	4	0	12
1	10	2	10
2	12	7	2
3	8	13	0

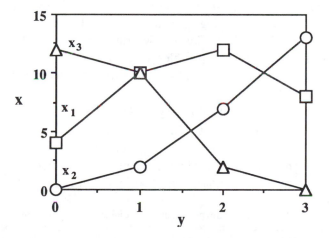

Fig. 6.11 - Graphs of three parameters in Table 6.9 against y.

In Fig. 6.11, we illustrate graphs of x_i against y. None of these show a linear trend. Probably the best variable to monitor is x_2, but even here the trend appears roughly quadratic or exponential and certainly cannot easily be used as a linear measure of y. Yet, buried in these measurements, there is likely to be some function that is linearly related to y.

In order to analyse these data further it is normal to centre each variable, that is subtract the mean from each variable. Providing we remember the means we can always go back to the original data. In early sections on PCA we discussed the importance of mean centring. The new dataset is given in Table 6.10. The mean of y is 1.5, so we should add this back on to any estimated of the value of y later on.

Table 6.10 - Mean centred data from Table 6.9

y'	x_1'	x_2'	x_3'
−1.5	−4.5	−5.5	5.0
−0.5	1.5	−3.5	3.0
0.5	3.5	1.5	−1.0
1.5	−0.5	7.5	−7.0

We want to find some connection between y' and the measured variables. Clearly using some function of all three variables is likely to more useful in determining a trend rather than using only one variable. The most obvious way of reducing the dataset is by PCA. This was described in Section 4.2, although the emphasis was on reducing two variables to one, rather than reducing three variables to one or two new variables.[1] The first and second principal components are given by

$$x_1'' \quad = \quad 0.176\, x_1 + 0.728\, x_2 - 0.663\, x_3 \qquad (6.18)$$
$$x_2'' \quad = \quad 0.984\, x_1 - 0.111\, x_2 + 0.139\, x_3 \qquad (6.19)$$

A graph of x_1'' against y is illustrated in Fig. 6.12. It shows a better linear trend than any of the individual variables, suggesting that there is a major underlying factor directly proportional to the concentration of compound added

---

[1] Calculational procedures when there are more than two variables are slightly complex, and it is normal to use an iterative method: the NIPALS algorithm is described in the appendix in case the reader is interested in reproducing these results on a spreadsheet or using a low level programming language, although most chemometrics and statistics software contains routines for PCA.

to the mixture. The equation for the best fit straight line (assuming no errors along the y axis) is given by

$$\hat{y}_{\{1\}} = 0.161\, x_1'' \tag{6.20}$$

We will produce a better model by selecting the second principal component, and can repeat this procedure to give

$$\hat{y}_{\{2\}} = 0.161\, x_1'' + 0.055\, x_2'' \tag{6.21}$$

where $\hat{y}_{\{p\}}$ is the estimate of y' using a p-component model. These estimates are tabulated in Table 6.11.

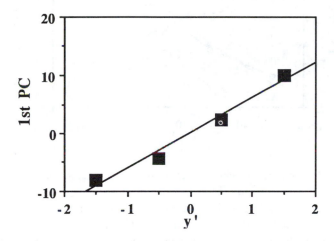

Fig. 6.12 - First principal component against y' as discussed for data of Table 6.11, together with best fit straight line.

Table 6.11 - Estimates of y'

y'	One PC model	Two PC model
−1.5	−1.303	−1.476
−0.5	−0.686	−0.560
0.5	0.381	0.554
1.5	1.608	1.481

These two estimates are illustrated in Fig. 6.13. The two component model is better than the one component model and provides fairly accurate information. This approach is termed *principal components regression* (PCR). In many typical spectroscopic cases there may be many hundred wavelengths and the problem of PCR is to fit a two or three component model to the variation in concentration of one or more compounds. Linear models are always used.

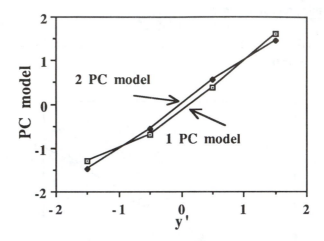

Fig. 6.13 - One and two component PCR models for calibration data.

One weakness of PCR is that the calculation of the principal components does not take into account the value of y. It is fortuitous that there is a good correlation between the principal components and y. It is possible to find examples where this correlation is not so good: this depends on the nature of the data. An alternative method, is called *partial least squares* (PLS). In this approach, the values of the dependent or y variable are taken into account during the calculation of the components. There are innumerable papers on the mathematical basis of PLS, and we have insufficient room in this text to go into much detail, but, essentially, instead of maximizing the variance (or sum of squares) of the first principal component, the covariance between y and the PLS component is maximized i.e.

$$c_p = \sum_{n=1}^{n=N} x''_{pn} \, y'_n / N \qquad (6.23)$$

for component p, where y'_n is the centred value of the y for sample n, there are N samples in the dataset and x''_{pn} is the PLS score of the centred pth component for this sample.

There are a variety of algorithms for performing PLS, one of which is given in the appendix, which can be very easily programmed into most simple programming languages or spreadsheets. The original algorithm, developed by S. Wold, is non-iterative if there is only one value of y. It is, however, important not to get confused between algorithms (generally reported in the chemometric literature) and actual use and interpretation of PLS components.

Table 6.12 - PLS components for data in Table 6.10

y'	x_1''	x_2''	$x_1'' + x_2''$
−1.5	−1.334	−0.142	−1.476
−0.5	−0.668	0.108	−0.560
0.5	0.409	0.145	0.554
1.5	1.592	−0.111	1.481

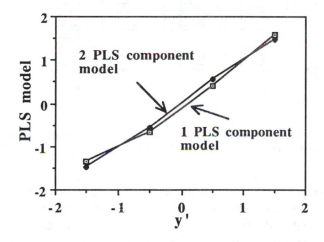

Fig. 6.14 - One and two component PLS models for calibration data.

The first two PLS components estimates of y are given in Table 6.12. An important feature of most PLS algorithms is that the coefficients are usually scaled so that the best estimate of y is merely the sum of scores for the individual components. Such scaling would be impossible in the case of PCA, since the method ignores the values of y which are only considered after dimensionality reduction. The first PLS component is closer in value to y than the first component produced by PCR, in this example. There is no significant difference when two components are taken into account as can be seen be comparing the last column of Table 6.12 to the two component model of Table 6.11. The two PLS models are illustrated in Fig. 6.14. Although, at first glance, there is very little difference between Figs 6.14 and 6.13, this is because of the scale we have chosen to draw them on: the influence of scaling is discussed in Section 4.2. An alternative way of comparing the one component PLS and PCR models is by plotting the graph of the absolute value of the error between y' and the two models (Fig. 6.15). The difference between successive y values in each experiment is 1, so an error of 0.2 is 20% of this difference. It can be seen from Fig. 6.15 that the absolute value of the error is smaller for PLS in all cases.

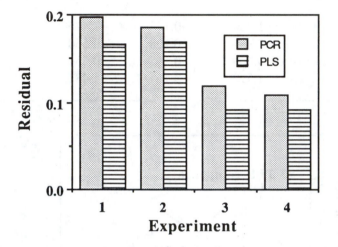

Fig. 6.15 - Absolute value of errors in one component PCR and PLS models
against experiment number (1 corresponds to a value to $y = 0$ and so on).

There are a huge number of papers contrasting PLS to PCR and related methods, but the main advantage of PLS is greater *interpretative* power. Because we use information on the value of y, the more significant PLS components are more likely to contain diagnostic information about the concentration of compounds in a mixture. Should we then always use PLS in preference to PCR? It is possible to find simulated and real examples where PLS significantly outperforms PCR, but, as we see in the example above, the difference is not always dramatic. There are no strong guidelines as to when PLS or PCR is preferred. An advantage of PLS is that it takes into account errors both in the y variable (e.g. concentrations) and in the x variables (e.g. spectroscopic measurements). In PCR it is assumed that there is no error in the y variable, which is often not the case. Of course if the y variable is very inaccurate PLS components will not be very meaningful and straight principal components might contain more diagnostic information.

The ideas behind PLS can be extended, and are of much interest to chemometricians. The major extension is called PLS2 to distinguish it from the method described above and in the appendix (sometimes called PLS1). The difference is that several y variables can be taken into account in the calculation. For example, we might be looking at the spectrum of a mixture of several components: instead of a single response, we have several values y_1 to y_m. The calculations become slightly more complex and we now produce m equations for each component, each equation being used to estimate each of the m responses. Another extension involves *multiblock PLS*. Spectra may depend on other factors such as pH (e.g. protonation equilibria), so we could have three blocks, namely, concentration(s), spectroscopic intensities and pH. Even more elaborate approaches can be built up calibrating several sets of measurements to each other. In a text of this size, we cannot do justice to this large battery of methods which are of active research interest to chemometricians and statisticians alike. One problem, though, is that very few good applications have been found to justify use of these elaborate methods of multiway calibration. Experimental design, signal deconvolution and repeatability of measurements need to be considered and the experimenter often spends more time overcoming these basic problems than obtaining sophisticated multiway data matrices, hence, many of the more recent approaches to multivariate calibration remain curiosities at present. This should in no way detract from the use of the more basic techniques, particularly PLS1, which are definitely superior to univariate calibration in many areas and which are increasingly incorporated into instrumental software.

One closely related topic is *structure property relationships* or *quantitative structure activity relationships* (QSAR). Because of the way this text is organized, multivariate calibration is discussed under multivariate signal processing, since some of the best established examples are in areas such as near IR spectroscopy, but a differently structured text might equally well introduce these topics in another context. Instead of the y variable being the concentration of a compound, it could be the pharmacological activity of a compound or the hardness of a material. These properties can be calibrated to structural or spectroscopic properties of a compound or material in exactly the same way as described above, and approaches such as PCR and PLS employed to connect these blocks of variables. The interested reader is referred to the article by Dunn in the bibliography for further reading in this area.

REFERENCES

Section 6.2
E.R.Malinowksi and D.Howery, *Factor Analysis in Chemistry*, Wiley, New York, 1980
R.A.Hearmon, J.H.Scrivens, K.R.Jennings and M.J.Farncombe, *Chemometrics Int. Lab. Systems*, **1**, 167 (1987)
E.R.Malinowksi, *J. Chemometrics*, **3**, 49 (1988)
S.Wold, *Technometrics*, **20**, 397 (1978)

W.Stone, *J. Roy. Stat. Soc. B*, **36**, 111 (1974)

Section 6.3
P.J.Gemperline, *J. Chemometrics*, **3**, 549 (1989)
R.J.Rummell, *Applied Factor Analysis*, Northwestern University Press, Evanston, Ill., 1970
P.K.Hopke, *Chemometrics Int. Lab. Systems*, **6**, 7 (1989)
B.G.M.Vandeginste, W.Denks and G.Kateman *Anal. Chim. Acta*, **173**, 253 (1985)
W.Windig, *Chemometrics Int. Lab. Systems*, **4**, 201 (1988)

Section 6.4
H.Martens and T. Næs, *Multivariate Calibration*, Wiley, Chichester, 1989
P.Geladi and B.R.Kowalski, *Anal. Chim. Acta*, **185**, 1 (1986)
I.R.Frank, *Chemometrics Int. Lab. Systems*, **1**, 233 (1987)
I.T.Joliffe, *Principal Components Analysis*, Springer-Verlag, Berlin, 1986
W.J.Dunn, III, *Chemometrics Int. Lab. Systems*, **6**, 181 (1989)
S.Wold, P.Geladi, K.H.Esbensen and J.Øhman, *J. Chemometrics*, **1**, 41 (1987)
A.Høskuldsson, *J. Chemometrics*, **2**, 211 (1988)

7

Pattern recognition

7.1 INTRODUCTION

The first investigators to describe themselves as chemometricians in the early 1970s were primarily interested in chemical pattern recognition. Approaches such as PCA, SIMCA, K-nearest neighbour classification and cluster analysis, which will be discussed in this chapter, were developed for the classification and comparison of samples principally from multivariate analytical measurements. Until the early 1980s, many people regarded chemometrics as synonymous with chemical pattern recognition. As this text has shown, there is a great deal more to chemometrics than pattern recognition. Indeed, there is little point classifying samples by their chemical fingerprints, for example, unless other aspects of the system are also considered. There is no point analysing data if the experiments are designed badly or if the signal intensities are meaningless. Pattern recognition does, however, play a central role in chemometrics: important medical, environmental, geochemical, biochemical and industrial conclusions can be made by interpreting chemical fingerprints.

There are a huge number of methods available for chemical pattern recognition, but most of these either aim to *classify* samples or to *interpret* the main features in a series of samples. Generally chromatographic or spectroscopic measurements are made on a series of samples, to give a *multivariate matrix* of objects (samples) and intensities (variables). Sometimes the concentrations of several compounds can be measured directly, such as by gas chromatography; in other cases these concentrations are buried within the signals such as in pyrolysis and near IR. Occasionally other measurements may be used: data such as colour, hardness and taste are also features of samples and related to chemical composition, origin etc. Physical properties like dipole moments and melting points can also be used.

In this chapter we discuss approaches for the interpretation of these multivariate chemical fingerprints. Previous chapters have discussed how to obtain such data and how to optimize and quantitate the measurement procedure. Even if the main objective of an experiment is to cluster a series of samples, pattern recognition cannot be considered in isolation to the rest of the measurement process. Until recently there has been a division in thinking between chemometricians interested primarily in pattern recognition and those interested in the measurement process. A major problem is the transfer and decoding of instrumental datasets between machines. Early pattern recognition studies were frequently performed off-line on large computers using FORTRAN or related subroutine libraries or statistical packages such as SAS. Datasets were not transferred automatically to the mainframe computer but manually typed in from printouts or traces obtained from laboratory instrumentation. The developers of on-line methods for signal processing and optimization such as Fourier transforms and Savitsky-Golay filters did not have ready access to these large FORTRAN and statistical subroutine libraries: only recently is it possible to run subsets of major statistical packages such as SAS and SPSS on micros. Frequently these micros are not linked directly to chemical instruments. Many laboratories do not have the expertise or cannot spare the time to develop software or hardware for efficient transfer of instrumental data to other computers running sophisticated pattern recognition software. Even if these problems are overcome, there are frequently difficulties with speed of transfer on a routine basis. Some instrument companies are beginning to overcome these problems and pattern recognition software is now available on certain types of spectrometers, but normally the software is restricted to a few well established approaches. Over the next few years there is, though, likely to be a growing awareness that chemometric methods are relevant to all stages of instrumental laboratory experiments, and this will be catalysed when a wider range of data-processing software is available on dedicated on-line computers linked to instrumentation. People whose main interest is in pattern recognition will sit down in front of a console of a spectrometer and find that they can easily improve the quality of data by changing acquisition parameters, filters, and so on; similarly practicing spectroscopists and chromatographers will discover a whole range of possibilities for processing sets of data after they have optimized, recorded, filtered, smoothed and integrated their signals.

7.2 PREPROCESSING DATASETS

The raw data for pattern recognition are normally a series of measurements on a series of samples. There is a huge battery of techniques for determining the relationship between these samples, but before using these methods it is important to prepare the data. This preliminary stage is often called *preprocessing*. Most chemometricians think of multivariate chemical data as matrices: each row represents a number of measurements of a single sample (or single experiment) and each column represents the measurements on a single variable (e.g. spectroscopic peak) over a set of experiments. This is illustrated in Fig. 7.1. Most multivariate methods for pattern recognition depend on analysing the relationships between measurements down each column (i.e. between each sample). In Section 4.2, we discussed the importance of *scaling* in chemometrics. Most methods of preprocessing involve scaling, and should be carefully compared and considered before subsequent multivariate pattern recognition. Methods for preprocessing can be divided into those that involve scaling along rows and and those that involve scaling down columns.

MULTIVARIATE MATRIX

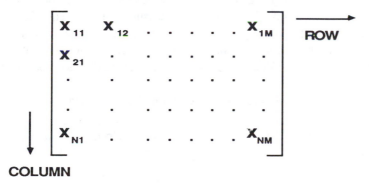

Fig. 7.1 - Multivariate data matrix - the N rows correspond to experiments, samples or observations, and the M columns to variables which normally are spectroscopic or chromatographic peak intensities or else the concentration of compounds.

It is usual first to scale data along rows (this is generally equivalent to scaling over a spectrum or chromatogram). There are several approaches.
(1) Leave the data unchanged. This is useful if absolute intensities are meaningful. For example, each variable may correspond to the absolute concentration of a compound in a sample. Normally the quantity of sample

used for each experiment must be identical (e.g. the experiments might be used to measure the concentration of chemicals in a given volume of blood from different patients). Great care must be taken to ensure that the instrumentation is truly quantitative. Absolute quantitation is rarely possible using techniques such as MS, for example.

(2) *Normalize* the data. This is a common transformation and involves scaling the variables for each experiment to a constant total. For example, if 30 mass spectral peaks are monitored for each sample, then the total mass spectral intensity, summed over all peaks, is set equal to 100 (or 1). This approach is useful if absolute quantitation is hard or unavailable. There can be difficulties with missing data and care must be taken to ensure that all relevant peaks are recorded for each experiment. The limitations with normalization are discussed below.

(3) Scaling the most intense peak for each experiment to 100. This is frequently performed automatically by instrumental software, but can be dangerous. There is rarely any physical reason to do this: if a set of experiments is performed in which the concentration of chemicals differs throughout a set of samples, the most intense peak is also likely to vary in each experiment.

(4) Using an external reference standard, and setting the value of this standard constant. For example, a compound of known concentration might be added to a mixture. The intensity of each peak relative to this standard is recorded. This approach sometimes provides a measure of absolute quantification (e.g. in NMR) but not always. In HPLC the intensity of a peak is dependent, in part, on the electronic absorption as well as the concentration, and compounds of different structures have different extinction coefficients.

Table 7.1 - Raw and normalized data

Sample	Raw data			Normalized data		
	x_1	x_2	x_3	x_1'	x_2'	x_3'
1	5	10	25	0.125	0.25	0.625
2	10	10	60	0.125	0.125	0.75

Although normalization appears to be the safest approach to scaling variables, there can be some difficulties. Table 7.1 illustrates a raw dataset and the corresponding normalized dataset in which the sum of the three intensities of the variables for each experiment equals 1. In most approaches to pattern recognition, the variability of the variables between samples is used in further calculation. In the raw data, variables x_1 and x_3 vary most, whereas x_2 is constant between the samples. When the data are normalized, the variability in $x_1^!$ is lost.

A compromise approach is called *selective normalization*. Certain variables are selected and the scaled data is calculated by

$$x_{in}^! = x_{in} / \sum_{j=1}^{j=q} x_{jn} \qquad (7.1)$$

where variables 1 to q are selected,[1] $q \leq I$ (the total number of variables measured), and x_{in} is the value of variable i for experiment n. If $q = I$ we normalize the entire dataset in the usual way. There are several criteria for selecting these variables, but a common one is to choose variables which have roughly similar means and standard deviations.

Scaling down the columns is the next stage. It is usual to subtract the mean from each of the columns, otherwise this is performed automatically by PCA (Section 4.2). It is possible, also, to *standardize* each variable, that is subtract the mean and divide by the standard deviation, so that

$$x_{in}^{''} = \frac{x_{in}^! - \bar{x}_i^!}{\sqrt{\sum_{m=1}^{m=N}(x_{im}^! - \bar{x}_i^!)^2/N}} \qquad (7.2)$$

where $x_{in}^!$ is the value of the reading for experiment n and variable i after it has been scaled along the rows; the total number of experiments is N and $\bar{x}_i^!$ is the mean of variable i over all N experiments.

The decision as to whether to standardize the variables or not depends on the nature of the problem. Consider a series of experiments that involve measuring the concentration of metabolites in a number of different species of organisms. We may wish to classify the organisms into different species according to their molecular fingerprints. Some compounds, such as the products of the major metabolic pathways, are likely to be present in high concentrations in all the organisms but the variation between species would not be terribly significant. Other compounds (secondary metabolites) may be present in far smaller quantities but be much more species specific. If the concentration of one of the secondary metabolites varies between 0 and 5 units, whereas the concentration of a less species specific compound varies between 100 and 140 units across the samples analysed, how can we ensure that the variation in concentration of the secondary metabolite (which may be highly significant) is suitably taken into account in the subsequent analysis? One approach might be to discard the information from the intense but less informative peak. This is called *variable selection*. However, we took the trouble to record several peaks, and unless we have a detailed knowledge of the system (which may involve spending more time studying the metabolic chemistry of the organisms) it is safest to include

---

[1] These do not need to be the first q variables measured; they may be any q variables.

information from all the variables. An alternative approach is to *weight* the variability of each peak equally: this means that the variability of the small peak is considered to be as significant as that of the large peak. The simplest way to do this is by standardization.

There is, though, a disadvantage to standardization. The precision of the least intense peaks is likely to be less than that of the most intense peaks. This is because of noise in the measurement process, as discussed in previous chapters. Therefore the procedure described in the paragraph above may amplify noise, destroying the advantage of standardizing the columns. There is no automatic solution to this dilemma, but it is possible to *weight* the variability for each peak by the transformation

$$x_{in}'' \quad = \quad (x_{in}' - \bar{x}_i')/\lambda_i \tag{7.3}$$

where λ_i is a weighting function. In Eq. (7.2) this is equal to the standard deviation of each variable. If it is known that a small peak is dominated by noise then the divisor can be large.

Most computer packages use, by default, either mean centred or standardized columns. Choice of the weighting function in Eq. (7.3) is often subjective and requires other knowledge about the nature of the errors in the data. There are no set and fast rules as to this choice except to urge all users of multivariate packages to consider very carefully the possible methods of preprocessing and scaling.

7.3 UNSUPERVISED PATTERN RECOGNITION

Unsupervised pattern recognition involves trying to determine relationships between objects (samples) without using prior information about these relationships. In supervised pattern recognition, certain objects are assumed to fall into predetermined groups or classes, which are used to produce class models.

The most widely used method for unsupervised pattern recognition in chemometrics is *cluster analysis*, which will be described in this section.

The raw data for cluster analysis consist of a number of objects and related measurements. In most situations measurements are quantitative, e.g. intensities of spectroscopic or chromatographic peaks, or concentrations of compounds. In other disciplines, such as biology or geology, *categorical variables* are often encountered: these are not quantitative variables and might either be the result of a simple negative / affirmative test (e.g. bacterial spot tests) or be represented by a range of possible states (e.g. colour or taste). Categorical measurements are rarer in chemistry and will not be considered in detail below.

The first step in cluster analysis is to scale the raw measurements as discussed in Section 7.2. To illustrate the various methods for cluster analysis, we will use a dataset consisting of five objects and eight measurements (Table 7.2). These data have not been normalized or standardized, but there is no difficulty repeating the analysis below using data scaled in different ways. For sake of brevity we restrict the discussion only to the raw data of Table 7.2.

Table 7.2 - Raw data for cluster analysis

Object	Variable							
	x_1	x_2	x_3	x_4	x_5	x_6	x_7	x_8
1	20	5	11	7	49	8	7	1
2	18	9	10	2	45	12	10	1
3	11	35	30	15	7	33	38	0
4	10	3	7	4	26	5	4	1
5	5	16	15	7	4	16	20	0

The next step is to determine the *similarity* between objects. The more similar objects are, the more likely they are to be related. For example we would expect the GC traces of the extract from two cheeses from the same area to be more similar than for two cheeses from different areas. There are a large number of measurements of similarity (or dissimilarity). The simplest ones are as follows.

(1) *Correlation coefficient.* This is defined in Eq. (3.1) where x is one variable and y is another variable. We redefine the correlation coefficient between object n and object m by

$$r_{nm} = \frac{\sum_{i=1}^{i=I}(x_{in}-\bar{x}_n)\cdot(x_{im}-\bar{x}_m)}{\sqrt{\sum_{i=1}^{i=I}(x_{in}-\bar{x}_n)^2\sum_{i=1}^{i=I}(x_{im}-\bar{x}_m)^2}} \qquad (7.4)$$

where I variables (eight in the example above) have been measured and x_{in} is the value of the ith variable for object n. The closer this coefficient is to 1 the more similar the two objects are. There is often some difficulty interpreting negative correlation coefficients, and, sometimes, the square of the correlation coefficient is used. In this text we only use the

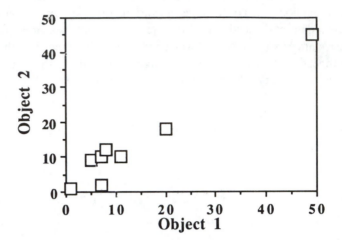

Fig. 7.2 - Graph of object 2 against object 1 (Table 7.2) giving a correlation
coefficient of 0.9761.

correlation coefficient defined by Eq. (7.4), which is called the *Pearson correlation coefficient*.[1] The correlation coefficient between object 1 and 2 is 0.9761: the measurements on the two variables are roughly linearly related (Fig. 7.2). We assume that negative correlation coefficients imply dissimilarity, although this is not always true.

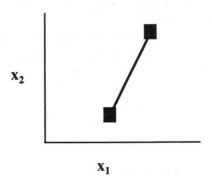

Fig. 7.3 - Euclidean distance between two points in two dimensional space.

[1]There are, in fact several different correlation coefficients, all with similar properties, that is the closer to ($\pm$)1, the more similar the objects.

(2) *Euclidean distance*. This is the *geometric distance* between two objects. We assume that each of the I variables is represented by an axis in I dimensional space. The distance for two objects in two dimensional space is illustrated in Fig. 7.3. The distance can be mathematically defined by

$$D_{nm} = \sqrt{\sum_{i=1}^{i=I}(x_{in}-x_{im})^2} \tag{7.5}$$

The closer this distance is to zero, the more similar objects are. Distances must be interpreted in the opposite way to correlation coefficients and are measures of dissimilarity.

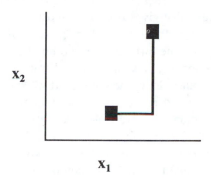

X_2

X_1

Fig. 7.4 - Manhattan or city block distance between two points in two
dimensional space.

(3) *Manhattan distance*. This is sometimes also called the city-block distance. It is the sum of the distance for each variable (Fig. 7.4) and is given by

$$D_{nm} = \sum_{i=1}^{i=I} |x_{in}-x_{im}| \tag{7.6}$$

The generalized *Minkowski distance* is given by

$$D_{nm} = \left(\sum_{i=1}^{i=I} |x_{in}-x_{im}|^p\right)^{1/p} \tag{7.7}$$

In the case of Eq. (7.5), $p = 2$, and for Eq. (7.6), $p = 1$. It is rare to use a value of $p > 2$.

The advantage of distances over correlation coefficients is that they are always positive. The disadvantage is that they can take any positive value and are not restricted to a range between 0 and +1. It is possible to convert distances to measures of similarity by dividing each variable by its *range*, $x_{i(\text{range})}$ (i.e. the difference between the maximum and minimum value of each variable, which, for example, equals $20 - 5 = 15$ for variable 1 in Table 7.2). The similarity measure for the Minkowski distance becomes

$$s_{nm} = 1 - \{\frac{1}{I}\sum_{i=1}^{i=I}[\ |x_{in}-x_{im}|\ /x_{i(\text{range})}\ ]^{p})\}^{1/p} \tag{7.8}$$

This similarity measure lies in the range 0 to 1, and can be interpreted in the same way as a correlation coefficient.

One difficulty with the measures of similarity mentioned above is that they do not take account of correlation between *variables*. This can be important in chemical studies. Consider recording the mass spectra of a series of samples each containing two compounds. One of the compounds may give rise to 10 mass spectral peaks, whereas another sample may give rise to only three mass spectral peaks. The compound that gives rise to the most mass spectral peaks will dominate the analysis since 10 out of 13 of the peaks will arise from this compound. Yet the reasons for the difference in the number of peaks might depend on factors such as molecular symmetry and not really have much relevance to the similarity or otherwise of the samples. If we knew this chemical information in advance we could reduce the dataset to two variables, but frequently this sort of information is not available. There are, however, methods for taking correlations between variables into account. The simplest approach is to use PCA to reduce the raw data. The number of significant principal components should, ideally, equal the number of compounds in the mixture or the number of significant factors that influence variability between samples, as discussed in Section 6.2. The same distances and correlations can be calculated between these new, reduced, variables.

Another approach is to use the *Mahalanobis distance*. This is defined by

$$D_{nm} = \sqrt{(\mathbf{x}_n-\mathbf{x}_m)\mathbf{C}^{-1}(\mathbf{x}_n-\mathbf{x}_m)'} \tag{7.9}$$

In this text we have largely avoided vectors except where essential. It is, however, hard to define this distance without resorting to vector or matrix notation. The vector $\mathbf{x}_n$ is the row vector $(x_{1n}, x_{2n}, x_{3n}, ...)$, so, for the dataset in Table 7.2, $\mathbf{x}_3 = (11, 10, 30...)$ and is eight columns long. As described in the appendix ' stands for transpose, so that the transpose of a row vector is a column vector. The matrix $\mathbf{C}$ is the *variance-covariance* matrix of the variables. Many of the more mathematically minded chemometricians use this matrix frequently in their computational procedures, but the emphasis of this book is not on computational algorithms but more on the results of using software. This matrix, however, is required in the definition of the Mahalanobis distance. In the example of Table 7.2, this matrix would contain eight rows and eight columns. The diagonals are the variance between the variables and the off-diagonal elements contain the covariance, so that

$$c_{ij} = \sum_{n=1}^{n=N}(x_{in}-\bar{x}_i).(x_{jn}-\bar{x}_j) / N \qquad (7.10)$$

where c_{ij} are the elements of the matrix.

Table 7.3 - Correlation coefficients and distances for samples in Table 7.2

Object	1	2	3	4	5
			Correlation coefficient		
1	1.000	0.976	−0.370	0.998	−0.338
2	0.976	1.000	−0.231	0.976	−0.199
3	−0.370	−0.231	1.000	−0.350	0.995
4	0.998	0.976	−0.350	1.000	−0.317
5	−0.338	−0.199	0.995	−0.317	1.000
			Euclidean distance		
1	0.00	9.33	68.97	26.00	51.20
2	9.33	0.00	62.96	23.64	45.45
3	68.97	62.96	0.00	63.06	36.17
4	26.00	23.64	63.06	0.00	33.60
5	51.20	45.45	36.17	33.60	0.00
			Manhattan distance		
1	0	23	165	48	97
2	23	0	154	51	86
3	165	154	0	149	86
4	48	51	149	0	79
5	97	86	86	79	0

The choice of distance or similarity is a difficult one. The method of preprocessing before calculating this similarity measure should also be taken into account. For example, in the data of Table 7.2, x_8 has a very small influence on distances or correlation coefficients. If the data are standardized down the columns, the variation in x_8 has a much larger influence. Which approach is used for preprocessing depends, of course, on the nature of measurement error and significance of the variables. The correlation coefficient is a good measure of *relative* similarity rather than *absolute* similarity. Two samples each consisting of one and the same compound alone that gives rise to several peaks will be exactly correlated. However, the distances will depend on the difference in concentration of the compound in the two samples.

As an example, we calculate the correlation coefficient, Euclidean distance and Manhattan distance for the five objects in Table 7.2: these are given in Table 7.3. From preliminary inspection it looks as if objects 1, 2 and 4 form a group, and objects 3 and 5 another group. This is very clear from the correlation coefficients, though not quite as obvious from the Euclidean distances: object 4 seems to be fairly dissimilar to objects 1 and 2, despite having a very high correlation with these objects. This is because many of the measurements on object 4 are roughly half that of object 1: distances take absolute differences into account. This grouping is less obvious using the Manhattan distance: object 5 seems about equidistant from all the other four objects, and it is especially interesting to note that the distance of object 5 from objects 2 and 3 is identical, in contrast to the correlation coefficients. The Manhattan distance appears to suggest a linear gradation of objects with object 3 at one end of the scale and object 1 at the other.

Some of the variables appear to be correlated, and an alternative approach is to reduce the variables to their principal components. The total sum of squares for the raw data (after the measurements for object have been mean centred) is 4222.80. The first two principal components can be calculated:[1] the sum of squares for the first principal component is 3477.27 and for the second principal component is 723.42, so that 99.5% of the variation in the data can be described by two principal components. Most of the tests described in Section 6.2 will select either a one or two PC model: below we employ two PCs. The five objects can be plotted in this new space (Fig. 7.5). The scores and loadings of the first two principal components are tabulated (Table 7.4).

It is interesting to note that objects 1, 2 and 4 have fairly similar values of x'_1, but the second principal component appears to group objects 1, 2 and 3 together.

[1]The principles behind PCA are discussed in Section 4.2. The interested reader can repeat these calculations using a variety of commercial chemometrics or statistical software. Alternatively it is easy to program a spreadsheet to perform PCA. The NIPALS algorithm is described in the appendix and can be readily implemented on most spreadsheets with a macrolanguage facility.

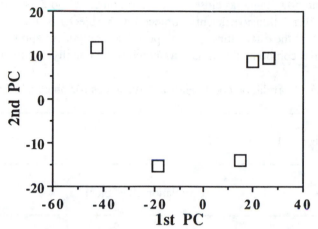

Fig. 7.5 - First and second principal components of data in Table 7.2.

Table 7.4 - First two principal components for data of Table 7.5

Loadings

	x_1	x_2	x_3	x_4	x_5	x_6	x_7	x_8
PC 1	0.14	−0.41	−0.28	−0.14	0.63	−0.34	−0.44	0.02
PC 2	0.35	0.33	0.28	0.13	0.69	0.33	0.31	0.01

Scores

Object	x_1'	x_2'
1	26.21	9.28
2	20.04	8.51
3	−42.68	11.56
4	14.96	−14.04
5	−18.53	−15.31

We can calculate the new distances and correlation coefficients (Table 7.5), and object 4 is no now longer as closely related to objects 1 and 2 as in the analysis above. The correlation coefficient of object 4 with object 1 is reduced from 0.99 to 0.46. One of the difficulties of using principal component approaches is that each principal component is usually assumed to be equally significant. In fact,

Table 7.5 - Correlation coefficients and distances for objects in Table 7.4

Object	1	2	3	4	5
			Correlation coefficient		
1	1.000	0.998	−0.823	0.459	−0.939
2	0.998	1.000	−0.786	0.404	−0.959
3	−0.823	−0.786	1.000	−0.883	0.578
4	0.459	0.404	−0.883	1.000	−0.126
5	−0.939	−0.959	0.578	−0.126	1.000
			Euclidean distance		
1	0.00	6.22	68.93	25.89	51.05
2	6.22	0.00	62.79	23.12	45.33
3	68.93	62.79	0.00	63.07	36.13
4	25.89	23.12	63.07	0.00	33.60
5	51.05	45.33	36.13	33.51	0.00
			Manhattan distance		
1	0.00	6.94	71.17	34.57	69.33
2	6.94	0.00	65.77	27.63	62.39
3	71.17	65.77	0.00	83.24	51.02
4	34.57	27.63	83.24	0.00	34.76
5	69.33	62.39	51.02	34.76	0.00

the first principal component accounts for 4.81 times[1] more information than the second principal component. It would be possible to weight the similarity measures to take this difference of significance into account.

We could, of course, use many other possible similarity measures, some of which have been discussed above.

Although it is possible to make some general comments about the similarity between objects and finish the calculation at this point, it is more usual to try to represent the information in a graphical form. In fact, it is hard to interpret large similarity matrices consisting of tens or hundreds of samples without some further data reduction. Again there are a vast number of possible ways of proceeding.

The most common method of cluster analysis is called *hierarchical agglomerative clustering*. There are several methods for hierarchical agglomerative cluster analysis. We will illustrate these using the Euclidean distances of Table 7.3.

The first step is normally to find the two objects of highest similarity. These are objects 1 and 2, since the distance is 9.33 (the smallest distance). These are joined together to form a cluster. The five objects are then reduced to four objects, namely "1,2", "3", "4" and "5". The new object "1,2" is also called a cluster. The next stage is to reduce the 5×5 similarity matrix to a 4×4 similarity matrix. There are several ways to do this. The most common are as follows.

(1) *Nearest neighbour* or *single linkage*. The distance between the new cluster and existing clusters or objects is the smallest distance between an object in the old cluster and the existing objects or clusters. So the distance between cluster "1,2" and object "3" is 62.96 rather than 68.97.

(2) *Furthest neighbour* or *complete linkage*. The greatest distance is used. In the example, the distance between "1,2" and "3" is 68.97.

(3) *Average linkage*. The average distance is taken. In the example above the average distance between "1,2" and "3" is 65.96. There are, in fact, two forms of average linkage. The most usual in chemistry is the *weighted pair group* method. If clusters A and B are joined together then the distance from cluster "AB" to "C" is the straight mean of the distance between "A" and "C" and "B" and "C". In the *unweighted pair group* method, the new distance is weighted by the number of elements in each cluster. If, for example. "A" consisted of two elements and "B" of one element, then the distance of "AB" to "C" would be given by

$$d_{\{AB\}C} = (2d_{AC} + d_{BC})$$ (7.11)

where d_{XY} is then distance between clusters "X" and "Y". It is important not to get confused with the terminology between weighted and unweighted pair

---

[1]This is the ratio of the sum of squares for principal components 1 and 2.

group methods: the weighting refers to the relative importance of individual members of a cluster, so that if a cluster of three objects has an equal influence on the calculation to a cluster of one object, the object in the latter cluster has three times the influence on the calculation, hence the reason for the definitions given above.

Table 7.6 - Single linkage (nearest neighbour) cluster analysis on Euclidean distance data of Table 7.3

Step 0

	1	2	3	4	5
1	0.00	**9.33**	68.97	26.00	51.20
2	**9.33**	0.00	62.96	23.64	45.45
3	68.97	62.96	0.00	63.06	36.17
4	26.00	23.64	63.06	0.00	33.60
5	51.20	45.45	36.17	33.60	0.00

Step 1

	12	3	4	5
12	0.00	62.96	**23.64**	45.45
3	62.96	0.00	63.06	36.17
4	**23.64**	63.06	0.00	33.60
5	45.45	36.17	33.60	0.00

Step 2

	124	3	5
124	0.00	62.96	**33.60**
3	62.96	0.00	36.17
5	**33.60**	36.17	0.00

Step 3

	1245	3
1245	0.00	**36.17**
3	**36.17**	0.00

Note : Smallest distance between objects and / or clusters denoted in bold in each step.

There are several other approaches, but the most common methods in chemometrics are the single linkage and the weighted pair group (average linkage) methods. Although there are numerous guidelines as to which method is best in particular situations, probably a useful approach is to use several linkage methods and see whether the overall results look fairly similar. If so, we have confidence in the overall results, and, if not, it is worth investigating whether there is any major reason why one method gives significantly different results.

Table 7.7 - Weighted pair group average linkage cluster analysis on data of Table 7.3

Step 0

	1	2	3	4	5
1	0.00	**9.33**	68.97	26.00	51.20
2	**9.33**	0.00	62.96	23.64	45.45
3	68.97	62.96	0.00	63.06	36.17
4	26.00	23.64	63.06	0.00	33.60
5	51.20	45.45	36.17	33.60	0.00

Step 1

	12	3	4	5
12	0.00	65.96	**24.82**	48.32
3	65.96	0.00	63.06	36.17
4	**24.82**	63.06	0.00	33.60
5	48.32	36.17	33.60	0.00

Step 2

	124	3	5
124	0.00	64.51	40.96
3	64.51	0.00	**36.17**
5	40.96	**36.17**	0.00

Step 3

	124	35
124	0.00	**52.74**
35	**52.74**	0.00

Note : Smallest distance between objects and / or clusters denoted in bold in each step.

The computations for single linkage and weighted pair group methods using the Euclidean distances of Table 7.3 are given in Tables 7.6 and 7.7. The two calculations come to somewhat different conclusions. Using the single linkage (nearest neighbour) criterion objects successfully join up in one large cluster. The most similar objects ("1,2") join first, then 4, 5 and 3. This phenomenon is often called chaining, whereby objects successfully join large clusters, with some outliers at the edges of the main cluster(s).

The results of this type of cluster analysis are normally represented graphically by a *dendrogram*. This is a diagram in which the closest objects are linked together first. Normally the vertical scale is distance (or correlation coefficient), and the objects are ordered along the horizontal scale so that there are no crossed lines. The dendrogram corresponding to Table 7.6 is given in Fig. 7.6 and that corresponding to Table 7.7 in Fig. 7.7.

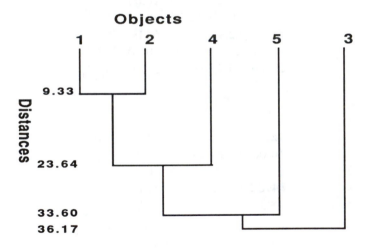

Fig. 7.6 - Dendrogram of data from Table 7.6 (single linkage nearest neighbour cluster analysis).

There are several other approaches for joining together objects: two of the most common are called the centroid method and Ward's method. For technical details of these methods, the reader is referred to the bibliography at the end of this chapter.

A little used alternative is *divisive hierarchical clustering*. In this approach the objects are assumed to form one cluster, which is then split into smaller clusters, and so on. The most usual method (that of MacNaughton) divides the clusters into two subclusters at each level, until ultimately the clusters consist of single objects. In most chemometrics texts divisive and agglomerative methods are regarded as completely different approaches. This is not entirely true. An interesting feature of dendrograms is that they represent a record of the steps in the algorithm used to cluster the objects. This distinguishes hierarchical cluster analysis from most other chemometric approaches, where the algorithm is merely a means to an end, and the steps in the calculation are of little interpretative value. Most early chemometricians were also developers of software and so did not readily distinguish algorithms from statistics. In fact, it can be shown, that many methods for hierarchical divisive clustering give identical results to corresponding methods for agglomerative clustering.

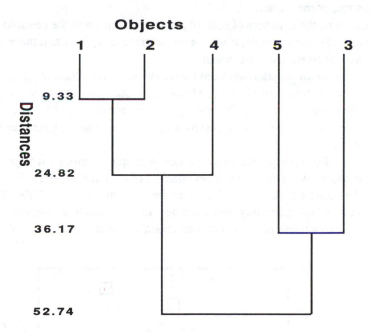

Fig. 7.7 - Dendrogram of data from Table 7.7 (weighted pair group average linkage cluster analysis).

In the methods discussed so far, there is no formal definition of a cluster. Certain objects are similar to each other, and some objects group or cluster together earlier in the process. In the example of Table 7.7, we do not formally define objects 1, 2 and 4 as belonging to one cluster and objects 3 and 5 as belonging to a second cluster. From the dendrogram of Fig. 7.7, it appears that 1, 2 and 4 form a group which is different to the group formed by 3 and 5. Each object is joined individually to other objects.

Non-hierarchical cluster analysis is another approach which aims to divide a set of objects into the most likely clusters. One method is called *optimization partitioning*. There are several variants. A common approach is as follows.

(1) Decide on the number of clusters (some methods involve a variable number of clusters, others involve searching for the optimal solution for a fixed number of clusters).

(2) Divide the data into k clusters (there are several methods for working out the initial "guesses"), so that, for example, if there are 20 objects in the dataset, and we want to determine three clusters, we might assign 10 objects to one cluster, 6 to a second cluster and 4 to a third cluster.

(3) Determine the centre of each of the k clusters. This is normally the geometric centroid of the cluster.

(4) Determine the distances of each object in each cluster to the centroid of each cluster. So, for example, if we are considering 20 objects and three clusters, there will be 60 distances in total.

(5) If the optimum solution has been found, the minimum distance of each object will be to the centroid of the cluster it is currently assigned to, and the method has converged. Otherwise, objects are reassigned to the clusters whose centroids they are closest to, and steps 3 to 5 are repeated for the new clusters.

In order to illustrate this we choose a new example. Although it is possible to use the data above, since the dataset consists of only five objects, the example would be rather uninteresting. Therefore we use a new dataset (Table 7.8). For simplicity, we consider only two variables. In practice, these variables may be principal components, or else multidimensional distances could be employed.

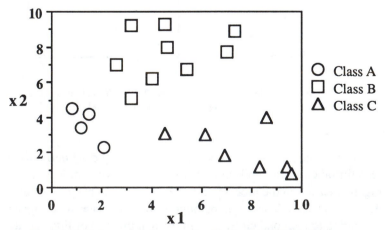

Fig. 7.8 - Initial guess of class membership for data of Table 7.8.

Table 7.8 - Data for non-hierarchical cluster analysis example

Object	x_1	x_2	Class (initial)	Class (final)
1	1.5	4.2	A	A
2	3.2	5.1	B	A
3	2.6	7.0	B	B
4	6.1	3.0	C	C
5	8.3	1.2	C	C
6	0.8	4.5	A	A
7	7.0	7.7	B	B
8	3.2	9.2	B	B
9	4.5	3.1	C	A
10	6.9	1.8	C	C
11	9.6	0.8	C	C
12	1.2	3.4	A	A
13	5.4	6.7	B	B
14	4.6	8.0	B	B
15	4.0	6.2	B	B
16	2.1	2.3	A	A
17	7.3	8.9	B	B
18	9.4	1.2	C	C
19	4.5	9.3	B	B
20	8.6	4.0	C	C

Note : Initial class is the initial guess of class membership; final class is the result of the clustering.

Let us assume that there are three groups (or classes) of object. A guess of the three classes (A, B and C) is illustrated in Fig. 7.8. The initial guess is given in Table 7.8. Four objects are in class A, nine in class B and seven in class C. The centroid of class A is ([1.5+0.8+1.2+2.1]/4, [4.2+4.5+3.4+2.3]/4) = (1.40, 3.60), of class B is (4.64, 7.57) and of class C is (7.63, 2.16). The distance of each object from the centroids of these original three classes is given in Table 7.9. Object 9 is closer to the centroid of class A than class C, so is reclassified. The class centroids are recalculated, and so on. The stable three class solution is given in the final column of Table 7.8. Two of the original objects were misclassified and are reassigned to class A. The new classes are illustrated in Fig. 7.9.

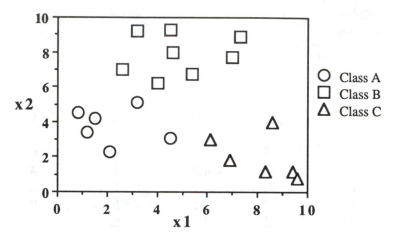

Fig. 7.9 - Final class membership for data of Table 7.8, using
non-hierarchical cluster analysis as described in the text.

It is slightly questionable as to how valuable this approach is for the simple example discussed above, but for large, multidimensional datasets, this and related methods can very usefully sort out patterns in the data. A large number of other methods for cluster analysis have been proposed by Massart and coworkers, to whose texts and papers the interested reader is referred.

Finally, mention should be made of a *minimal spanning tree*. This is one of the earliest examples of chemometric pattern recognition, and is conceptually very simple. The "tree" is constructed as follows.

(1) The distances between all the objects are calculated.

(2) The smallest distance is selected and the two objects with this distance are joined up.

(3) The next smallest distance is then selected and the relevant objects joined up unless a closed circuit is formed in which case step 4 is performed. If it does not form a close circuit step 3 is repeated until all objects have been joined to at least one other object, in which case step 5 is performed.

(4) The current distance is eliminated from the distances under consideration, and step 3 is repeated on the next smallest distance.

(5) The largest inter-object distance is taken as the division between two clusters; generally for three or more clusters the next largest distance is taken, and so on.

Table 7.9 - Distances of objects from centroids of the three initial classes in Table 7.8

Object	Class A	Class B	Class C	Most favoured class
1	0.61	4.61	6.46	A
2	2.34	2.86	5.32	A
3	3.61	2.12	6.98	B
4	4.74	4.79	1.75	C
5	7.31	7.34	1.17	C
6	1.08	4.92	7.22	A
7	6.94	2.36	5.58	B
8	5.88	2.18	8.32	B
9	3.14	4.46	3.27	A
10	5.79	6.19	0.81	C
11	8.66	8.39	2.39	C
12	0.28	5.41	6.55	A
13	5.06	1.15	5.06	B
14	5.44	0.44	6.58	B
15	3.68	1.51	5.43	B
16	1.48	5.85	5.53	A
17	7.93	2.97	6.75	B
18	8.35	7.95	2.01	C
19	6.49	1.74	7.80	B
20	4.76	9.99	4.61	C

We will illustrate this calculation using the data of Table 7.10 in Fig. 7.10. Some interesting points to note are as follows. Object 3 is joined to three other objects, whereas objects 4, 5 and 6 are only joined to a single other object. The largest distance in the tree is between objects 1 and 2, so we could divide the data into two clusters, namely, (1) objects 1 and 4, and (2) objects 2, 3, 5 and 6. It is easy, for example, to see that object 2 is closer to object 5 than it is to object 1, but is not joined up in the tree because this would create a closed circuit.

Statisticians have developed a huge literature on cluster analysis, which is extensively employ in biology (e.g. taxonomy) and geology. Large and specialized software packages such as CLUSTAN are available. Extensive clustering options are available in packages such as SAS. Many approaches have yet to be applied to chemical data.

Table 7.10 - Data for minimal spanning tree

Object	x_1	x_2
1	3.7	7.2
2	6.8	3.1
3	6.9	2.6
4	7.6	8.4
5	8.4	1.2
6	4.5	0.9

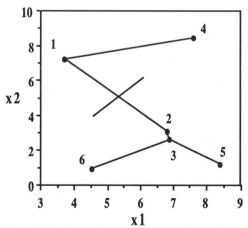

Fig. 7.10 - Minimal spanning tree corresponding to the objects in Table 7.10, as described in the text. The longest distance, namely that between objects 1 and 2, is indicated.

There are no strong guidelines as to the most suitable approach for unsupervised pattern recognition in chemistry. Probably the best tactics are to try several methods and see whether the results are roughly similar. Some of the most useful examples of clustering methods have been applied to large spectroscopic libraries. In many other areas of chemometrics, however, supervised methods of pattern recognition, described below, have found greater use.

7.4 SUPERVISED PATTERN RECOGNITION: HARD MODELLING

In supervised pattern recognition objects are classified into groups (or classes or clusters) with predetermined models for the class. These approaches differ from unsupervised methods such as cluster analysis where there is no prior class model. Many methods for unsupervised pattern recognition do not even attempt to classify objects, but merely reduce the complex raw data into a means of visualizing relationships between objects, such as a dendrogram.

The aim of hard modelling, a form of supervised pattern recognition, is to classify an object uniquely into a number of predetermined classes. There is a great deal of classical statistical literature in this area. In many traditional applications, classes are well defined groups. For example, each class may represent a different species. The data might consist of a number of measurements on a set of plants from three species: these measurements might be of petal length, stem length etc. Multivariate statistics can be employed to find the most diagnostic features (or combination of features) that distinguish one species from another. There is no ambiguity in the classification of a plant: it cannot be a member of two species at the same time, so statistical approaches were developed to provide categorical answers about class membership from taxonomic and other measurements.

These approaches can be applied in chemistry to classify samples. Examples of classification problems might be determining whether a double bond is *cis* or *trans* from a set of spectroscopic measurements, or determining whether an extract comes from a tumorous or non-tumorous biological tissue from its chromatographic profile. The classical approach to pattern recognition in chemistry is to select a *training set*. This consists of a number of objects of known or assumed class. Sometimes the groups are obvious. For example, it is easy to measure the IR spectra of a series of commercially available *cis* and *trans* alkenes. We are absolutely sure which of the training set belongs to which class. In other cases, the classification may not be so well defined, but cluster analysis and related methods discussed above may be used to determine the number and membership of the main groups, which can then be used as training sets to determine class models. This latter approach may be used, for example, when dealing with large chemical structural databases where the major groups are not clear from a first inspection. The training set is used to set up a *class model* in which features can be used to classify new (unknown) objects. This means, for example, we could use the information obtained from the spectra of know *cis* and *trans* alkenes to determine the stereochemistry of an unknown compound.

We will consider as an example, the data of Table 7.11, in which two measurements are taken on 13 objects, seven of which belong to one class and six to another class. The graph of the data is given in Fig. 7.11. Class B is elongated vertically in this example, and most of the readings are in the top right-hand corner. Class A, however, is elongated horizontally, and tends to occupy a crescent at the bottom of the graph. Can we calculate a numerical value that distinguishes the two classes?

Table 7.11 - Example for supervised pattern recognition.

Object	x_1	x_2	Class
1	0.939	−4.847	A
2	−5.128	−1.255	A
3	−2.983	−2.149	A
4	−1.964	−1.019	A
5	0.151	−3.461	A
6	2.066	−2.863	A
7	1.998	−0.641	A
8	1.857	5.068	B
9	−0.389	2.187	B
10	1.621	3.058	B
11	1.153	3.888	B
12	0.868	1.147	B
13	−0.189	0.887	B

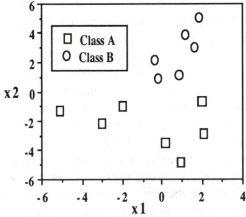

Fig. 7.11 - Data of Table 7.11, separated into classes

One criterion might be to say that if x_2 is less than 0.0 the object belongs to class A, otherwise to class B. But this criterion only uses the information from one variable, and is chosen quite arbitrarily. In a very simple example like this one it is certainly possible to determine these rules by eye, but in practice far more sophisticated approaches are required. Normally the aim of hard modelling is to produce a *discriminant function*. If an object is one side of the discriminant function it is assigned to one class, and if the other side to another class (for two classes). An ideal discriminant function will divide the data into two discrete classes.

There are many ways of computing this function, but the classical approach, devised by the statistician R.A.Fisher, is called *linear discriminant analysis* (LDA). A direction is found that passes through the data. This direction is a straight line, and is somewhat analogous to a principal component, but a different criterion is used to find this line. Instead of maximizing the sum of squares of the residuals (see Section 4.2), the ratio of the between groups to the within groups square error is maximized. For the case where there are only two classes, A and B, and the data is reduced to a single new variable, x_1'. The vairable R_{AB} is a maximum, where,

$$R_{AB} = \frac{(\bar{x}_{1A}'-\bar{x}_{1B}')^2}{\sum_{n=1}^{n=N}(x_{1n}'-\bar{x}_1')^2} \qquad (7.12)$$

$\bar{x}_{1C}'$ is the mean of the transformed variable for class C, $\bar{x}_1'$ is the mean of the transformed variable over the entire dataset, and x_{1n}' is the value of the transformed variable for object (or experiment or sample) n. One way of looking at this transformation is as a *projection* onto a straight line. For the data of Table 7.11, the equation for the transformation onto this straight line is given by

$$x_1' = 0.223 x_1 + 0.975 x_2 \qquad (7.13)$$

so that the straight line is

$$\hat{x}_2 = 0.975 / 0.223 \,\hat{x}_1 = 4.41 \,\hat{x}_1 \qquad (7.14)$$

The interested reader is referred to the appendix for further details of how to calculate this function.

Note that the data of Table 7.11 are chosen so that the overall mean of both raw variables is 0. If this is not so, the data should be mean centred in advance. We can plot the straight line defined by Eq. (7.13) (Fig. 7.12) which defines a *discriminant function* or the first *canonical variate*. It is constructive to compare this to the first principal component, given by

$$\hat{x}_2 = 2.784 \,\hat{x}_1 \qquad (7.15)$$

or

$$x_1' = 0.338 x_1 + 0.941 x_2 \qquad (7.16)$$

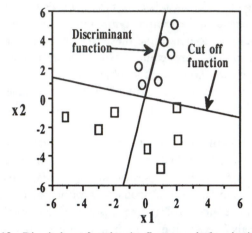

Fig. 7.12 - Discriminant function (or first canonical variate) for data of
Table 7.11; the cut-off function that ideally separates the data into two
classes and is at right angles to the discriminant function is also
illustrated.

Table 7.12 - First principal component and discriminant function for data of
Table 7.11

Object	Principal component	Discriminant function	Class
1	−4.24	−4.52	A
2	−2.91	−2.37	A
3	−3.03	−2.76	A
4	−1.62	−1.43	A
5	−3.20	−3.34	A
6	−2.00	−2.33	A
7	0.07	−0.18	A
8	5.39	5.35	B
9	1.93	2.05	B
10	3.42	3.34	B
11	4.04	4.05	B
12	1.37	1.31	B
13	0.77	0.82	B

The scores for the first principal component and the first canonical variate (discriminant function) are tabulated (Table 7.12). The principal component does not distinguish the two classes quite as well: the distance between objects 7 and 13, for example, is closer in the principal component solution than the canonical variate solution. Object 7 looks much closer to class B than class A in the principal component solution, whereas the "gap" between the classes is bigger in the canonical variate solution. Principal components methods are not provided with information as to class membership, so it is fortuitous that the principal component solution manages to partially separate the classes in this example.

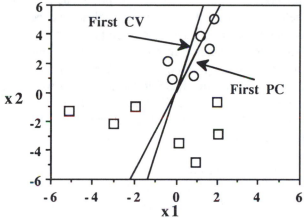

Fig. 7.13 - First principal component and first canonical variate for data of Table 7.11.

The projections of the data onto the first principal component and discriminant function are given in Fig. 7.13. In many statistical texts a *cut-off* value for the discriminant function (canonical variate) is determined. This is defined by

$$c'_{AB} \quad = \quad (\bar{x}'_{1A} - \bar{x}'_{1B})/2 \qquad\qquad (7.17)$$

i.e. the mean of the means of the discriminant function for the two classes, which, in the example above is given by 0.32. This cut-off can be visualized as dividing the data into two areas, one corresponding to class A and the other to class B, and may be represented by a line at right angles to the discriminant function (Fig. 7.12). Therefore readings with a value of x'_1 less than 0.32 are classified into class A, and above this value into class B.

Discriminant functions are not always as clean as in the example above. Sometimes there is a little overlap between two classes: it is not possible to guarantee that a cut-off value will exactly separate classes. However, discriminant functions use knowledge available to the chemometrician. In the example above we did not use this knowledge in constructing a principal component model, and it is not at all obvious that there are two groups and also where one group starts and the other ends. Which class does a sample with a score of 0.5 belong to? There is no clear answer from PCA, whereas CVA (canonical variates analysis) will classify it into class B.

Normally, after a model has been set up, a *test set* of samples is used to test how good the model is. For example, 13 compounds, seven with a *cis* double bond, and six with a *trans* double bond might have been used to set up the model of Table 7.12. We then calculate the discriminant function for a 14th compound of known stereochemistry and see which class it is assigned to. If several of these tests work correctly, we are reasonably confident that the whole procedure is behaving well. The occasional misclassification is to be expected. A good way of sorting out these misclassifications is to employ two completely different analytical procedures: for example, discriminant analysis could be performed on near IR and pyrolysis mass spectra of compounds and the results compared. If both approaches classify a compound into the same class we are very confident that our assignments are correct. If there is some ambiguity in classification it may be worth using completely different approaches for determining the type of compound. Discriminant analysis can be used as a quick screening that will sort out ambiguities.

There is no reason, of course, to restrict CVA to two classes or two variables. The calculations become somewhat computationally complex for a larger number of parameters. In many cases two or three canonical variates are calculated, and projections onto these lines are computed in a superficially similar way as in PCA. There is one major difference, though, and that is canonical variates do not need to be at right angles to each other, unlike principal components. This means that distances (space) will be distorted from the normal Euclidean space. A major limitation of CVA is that most computational algorithms will not work if the number of variables minus the number of groups is less than the number of objects (samples). Ways of getting around this exercise the minds of some statisticians, but one approach involves reducing the number of variables by an approach such as PCA first, and then performing CVA on the reduced dataset. This problem can be quite serious in cases such as pyrolysis where a large number of measurements might be made on a series of samples. Should some of the measurements be discarded for the sake of performing CVA? Fortunately there are several ways to solve this problem: the original developers of computational algorithms did not take into account situations where measurements are plentiful compared with samples, and so they wrote this limitation into the software. Since LDA has not been greatly used by chemometricians it is possible that readily available computational methods have not yet been developed to overcome this limitation. To the user of CVA it is important to realize that most generally accessible statistical methods for CVA are limited in this respect, and to assess carefully how much time is available for developing computational approaches to overcome this problem or whether alternative data analytical approaches will reach a satisfactory answer much more quickly.

A different approach to finding discriminant functions is called a *linear learning machine* and is probably one of the first applications of pattern recognition to chemistry, published in the late 1960s. We will restrict the discussion below to cases where the aim is to separate two classes. The aim of this method is to devise a linear function of the form

$$f \quad = \quad \sum_{i=1}^{i=I} \alpha_i x_i \qquad\qquad (7.18)$$

for each sample (object), where a positive value of f classifies the sample into one class, and a negative value into another class, and I measurements are taken on each sample. There are several possible ways of determining this function,

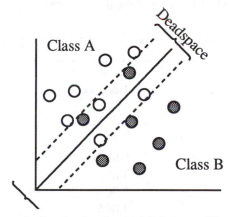

Fig. 7.14 - Example of a situation in which it is impossible to define a linear discriminant function that unambiguously separates two classes, and of the use of a deadspace between classes if class membership is undefined.

and the methods are normally developed as algorithms rather than using statistical criteria. Most approaches involving guessing an initial function, testing one sample and then working out whether the function correctly classifies this sample. If it does, another sample in the training set is tested. If it does not, the function is adjusted to classify correctly this sample and then a new sample is tested. There are, of course, several methods for adjusting the function, beyond the scope of this text. However, the method depends on the order in which the samples are tested and also works effectively only if there is an unambiguous linear discrimination between classes. In the case of Fig. 7.14, it is not possible to determine a function of the form of Eq. (7.18) that will correctly work in all cases. This limitation can be overcome by defining a "dead zone" either side of the discriminant function in which objects cannot be unambiguously classified. Many other sophistications have been developed. These include quadratic discriminant functions, and other approaches such as simplex optimization to determining a function of the form of Eq. (7.18). One of the problems is that the nature of the resultant function is dependent on the algorithm used. This is in contrast to most of the other methods described in this chapter.

Finally, we discuss the *K nearest neighbour* (KNN) method. The Euclidean distance of an unknown to the members of the training set is calculated. The object is assigned to the class to which the majority of the K nearest neighbours belong. Generally K is an odd number (e.g. 3 or 5) so that classification in unambiguous. Consider an object with measurements (x_1, x_2) = (0.456, −0.381). Does this object belong to class A or B of Table 7.11? As shown in Fig. 7.15 it is quite difficult to decide on purely visual inspection. The distances are tabulated in Table 7.13. Two of the three nearest neighbours (objects 12 and

Table 7.13 - Distances of unknown object from objects of Table 7.11

Object	Distance	Class
1	4.492	A
2	5.652	A
3	3.867	A
4	2.503	A
5	3.095	A
6	2.958	A
7	1.563	A
8	5.626	B
9	2.704	B
10	3.631	B
11	4.326	B
12	1.582	B
13	1.422	B

13) belong to class B, so using a 3 nearest neighbour classification rule, we assign it to class B; this is also true for a 5 nearest neighbour classification rule, in this case. It is constructive to compare the KNN classification to the canonical

variates classification. Using Eq. (7.13), the value of the discriminant function is −0.270, so the sample is assigned to class A rather than class B (indeed it is closer to the mean of class A than object 7 is).

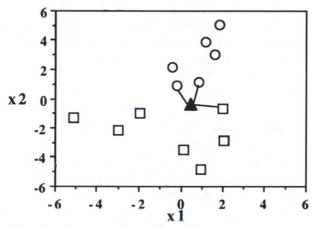

Fig. 7.15 - Illustration of the K Nearest Neighbour method. The distance of the unknown object (co-ordinates [0.456, -0.381]) to three closest members of the existing two classes is measured.

There have been several modifications to the KNN approach in chemistry, but these are omitted for reason of space. It is probably unwise to become too wedded to any one approach for supervised pattern recognition, and, rather than rely on elaborate computational and algorithmic modifications to any one individual method, it is best to try out a variety of methods and measurement procedures before coming to conclusions.

7.5   SUPERVISED   PATTERN   RECOGNITION: SOFT MODELLING

One of the most widely publicized methods of chemical pattern recognition is SIMCA (called, *inter alia*, soft independent modelling of class analogy). This was developed in the 1970s by one of the main proponents of chemometrics, S. Wold. For many years, SIMCA was almost synonymous with chemometrics. Innumerable papers have been published applying this method to classifying chemical data. Probably the reason for the success of the method is that the proponents developed user-friendly software around this method. SIMCA was one of the earliest available micro-based chemometrics software packages, and so reached the laboratories of many less technically minded users of chemometric methods. Approaches such as canonical variates and cluster analysis were largely implemented on specialized statistical software packages, and, although of considerable interest to the practitioners, not in such a user-friendly form as SIMCA. Therefore the early literature and applications of SIMCA overshadow that of nearly all other chemometric methods for pattern recognition. This does not necessarily imply that this approach is automatically better than any of the other methods available for classifying groups of samples, but is an example of good salesmanship and attention paid to software development. As it becomes increasingly easy to develop software, and as computer memory and time limitations are less severe, and microcomputers become ever more easy to program and use, it is unlikely that future methods will depend so heavily on selling software. However, the SIMCA method can be implemented in most packages and on most computer systems and is a major tool of interest to the chemometrician.

The methods described in Section 7.4 rely on giving a categorical answer about class membership. Each sample is assigned to one single group, even if there are difficulties with the assignment. Sometimes, though, there is insufficient evidence to unambiguously classify a sample. We might want an answer such as there is a 90% probability that the sample belongs to group A. Typically we may not have enough evidence to determine unambiguously, for example, whether a sample of river water comes from one river or another, from chromatographic measurements. In SIMCA each class is modelled independently of the other classes. One major advantage of SIMCA is that it is possible to add classes to the calculation without changing the overall model. For example, we might determine a set of measurements on ketones and another set of measurements on esters, and use this information to determine whether an unknown is an ester or a ketone. Several months later we might decide to extend the database to amides. If we use hard modelling we would need to completely change the calculations and recompute canonical variates, clusters, KNN criteria or whatever. In the case of soft modelling we just add another, independent, class.

We will illustrate the SIMCA method using the data of Table 7.11. We will model both classes by a single principal component, and find that the best fit models are

$$x_{1(A)}' = 0.967 (x_1 + 0.703) - 0.256 (x_2 + 2.319) \qquad (7.19)$$

and

$$x_{1(B)}' = 0.434 (x_1 - 0.820) + 0.901 (x_2 - 2.706) \qquad (7.20)$$

where $x_{1(C)}'$ is the value of the score of the first principal component fitted to the data of class C. The reader should note that -0.703, for example, is the mean of x_1 over class A and is subtracted from x_1 prior to constructing the principal component model.[1] The models of Eqs (7.19) and (7.20) give the scores for a one principal component model, and are tabulated in Table 7.14.

Table 7.14 - Scores for principal component models of Eqs (7.19) and (7.20)

Object	Score: model class A	Score: model class B	Class
1	2.234	−6.754	A
2	−4.550	−6.148	A
3	−2.248	−6.023	A
4	−1.551	−4.564	A
5	1.117	−5.847	A
6	2.816	−4.478	A
7	2.182	−2.505	A
8	0.587	2.578	B
9	−1.037	−0.992	B
10	0.684	0.665	B
11	0.020	1.210	B
12	0.445	−1.384	B
13	−0.511	−2.077	B

Note : The scores are for the individual class models, and, therefore differ from the scores given in Table 7.12 which are for the entire dataset.

The next step is to see how well an object fits either class. The 13 known objects can be fitted to either class model as follows.

$$\hat{x}_{in\{C\}} = \bar{x}_{i\{C\}} + \sum_{p=1}^{p=P} x_{pn\{C\}}' \alpha_{ip\{C\}} \qquad (7.21)$$

where $\hat{x}_{in\{C\}}$ is the estimated value of the ith variable for the nth object using the model of class C, $\bar{x}_{i\{C\}}$ is the mean of variable i over class C, $x_{pn\{A\}}$ is the score for the pth principal component and object n, $\alpha_{ip\{C\}}$ is the corresponding loading and we use P principal components in the model. In the example discussed in this section $P=1$, i.e. we use a one component model. In most real cases there will be several (perhaps even several hundred) variables, and so each class could be modelled by several principal components. The value of P is chosen using

---

[1] In most of the examples in this text the datasets have been mean centred. Because each class has a different mean it is necessary to include terms for the mean in the class model. The mathematics would simplify considerably if both classes had identical means, but this is very rarely the case in practice.

approaches discussed in Section 6.2. Most packages use cross-validation because the originators of SIMCA were strong proponents of this method for determining the optimum number of components in a model, since cross-validation has the advantage of selecting those components with interpretative value, which is important when trying to classify objects by their differences. It is possible to fit different classes to models containing different numbers of components, so one class might be best modelled by three components and another by two, for example.

As an example we fit object 4 to class model A^1: $\hat{x}_{14\{A\}} = -0.703 - 1.551 \times 0.967 = -2.028$; $\hat{x}_{24\{A\}} = -2.319 - 1.551 \times -0.256 = -1.923$. These compare to the observed values of $x_{14} = -1.964$ and $x_{24} = -1.019$ (Table 7.10). We can also fit the object to the model of the other class (B), and expect this to be a worse fit. The classes may be visualized graphically, by a plot of the principal component models for each class (Fig. 7.16). The difference between the observed and predicted values for each variable can be converted to geometric distances in the normal manner, so the error for object n fitted to class model C is

$$\varepsilon_{n\{C\}} = \sqrt{\sum_{i=1}^{i=I}(x_{in}-\hat{x}_{in\{C\}})^2/(I-P)} \qquad (7.22)$$

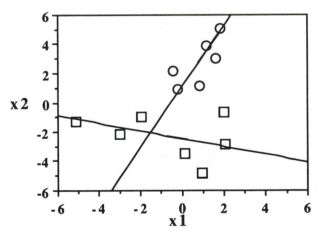

Fig. 7.16 - One principal component models for classes A and B (Table 7.14).

where there are I variables in the original data. It is, therefore, possible to fit objects of class A both to the model of class A and to the model of class B, and

---

[1]Use the score for model A, object 4 in Table 7.13 and coefficients for the first principal component in Eq. (7.19).

compare how well they obey each class model. The errors are tabulated (Table 7.15):

Table 7.15 - Root mean square errors when data of Table 7.11 are fitted to class models A and B

Object	Class model A	Class model B	Class
1	2.024	3.383	A
2	0.102	3.642	A
3	0.417	1.322	A
4	0.934	0.893	A
5	0.886	2.071	A
6	0.182	3.537	A
7	2.313	2.512	A
8	7.796	0.090	B
9	4.441	0.865	B
10	5.796	0.569	B
11	6.479	0.213	B
12	3.757	0.719	B
13	3.236	0.120	B

There are some interesting points to note. For example, object 4 appears closer to class B than class A. It is possible that some objects are misclassified, and this approach can help us identify these objects: it is possible to make a mistake and mislabel a sample. Another possible reason for misclassification is if the true class is not always known with certainty. However, a major advantage of soft modelling is that we do not force objects into discrete classes. Object 8 is clearly much closer to class B than class A, whilst object 6 is much closer to class A than class B. Object 4 could fit into both classes quite easily. An interesting object is number 1 which is distant from both classes. Sometimes there are samples that cannot be classified into any predefined groupings. We might be analyzing cheeses from two regions. What happens if we are given a single cheese that belongs to neither region? Using hard modelling the sample will almost inevitably be classified as belonging to one or other of the two groupings, but soft modelling allows us to detect *outliers*, that is samples that belong to none of the predefined groups.

A great deal more information can be obtained from SIMCA. The interested reader is referred to the references at the end of this chapter for a comprehensive description of these elaborations, but some of the key statistics are discussed below.

It is common to calculate the *residual standard deviation* of a class, defined by

$$R_{\{C\}} = \sqrt{\sum_{n=1}^{n=N_C}\varepsilon^2_{n\{C\}}/(N_C-P-1)} \tag{7.23}$$

where there are N_C members of class C (numbered from 1 to N_C). For class A this standard deviation is 1.505 and class B it is 0.644. Inspection of the graphs of classes A and B does, indeed, suggest that A is more spread out and inhomogeneous. The smaller this value, the "tighter" and better defined the class. Another statistic is the residual standard deviation when the data of one class is fitted to that of another class, defined by

$$R_{\{C|D\}} = \sqrt{\sum_{n=1}^{n=N_C}\varepsilon^2_{n\{D\}}/N_C} \tag{7.24}$$

where the data of class C are fitted to a model of class D. The distance between two classes is given by

$$d_{CD} = \sqrt{(R^2_{\{C|D\}}+R^2_{\{D|C\}})/(R^2_{\{C\}}+R^2_{\{D\}})} - 1 \tag{7.25}$$

This measure can be interpreted quantitatively: a value smaller than about 1.0 corresponds to very small class differences, and less than 0.5 negligible evidence for difference between the two classes. In our example this difference is 2.73, which strongly indicates differences between these classes. It is also possible to define the distance of an object from a class model. For the *training set* that is used to set up the class models, this distance is may be defined by

$$d_{n\{C\}} = \varepsilon_{n\{C\}}\sqrt{N_C/(N_C-P-1)} \tag{7.26}$$

so the distance is slightly greater than the errors in Table 7.15. For *test* samples (i.e. new objects that were not used to set up class models) the straight error is used without any scaling. These distances can be viewed graphically for pairs of classes – Fig. 7.17 illustrates a class plot for the data in this section. Another use of class distances is to try to classify an unknown into existing classes. Consider the object with $(x_1, x_2) = (0.456, -0.381)$ discussed in Section 7.4. The fit to class A is $(\hat{x}_{1\{A\}}, \hat{x}_{2\{A\}}) = (-0.099, -2.669)$ and to class B is $(\hat{x}_{1\{B\}}, \hat{x}_{2\{B\}}) = (-0.457, 0.058)$.[1] Therefore the distance from class A is 2.170 and class B is 1.013. This object could fit into either class but seems closer to class B, as illustrated in Fig. 7.16. It is possible to calculate the significance of the distance of an object to a class using an *F-test* (see Section 2.5 for more detailed

[1]See Eqs (7.19), (7.20) and (7.21) for details of the calculation.

discussion), using the ratio $d_{n\{C\}}^2 / R_{\{C\}}^2$, and so derive probabilities that a given object is the member of a class.

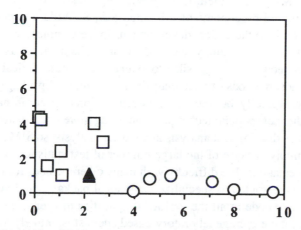

Fig. 7.17 - SIMCA class plot for data discussed in the text. The horizontal axis represents the distance from class B, and the vertical axis distance from class A (see Eq. (7.26)). The symbols are defined in Fig. 7.15.

A final piece of information often of interest is the *modelling power* of a variable. This can be very relevant to the chemist. For example, an experiment might be performed to measure the concentration of 20 chemicals in two species of bacteria. Some chemicals will be more species specific, whereas others might contain very little diagnostic information. It is more efficient to measure the concentration of the most diagnostic variables, particularly if each measurement is expensive. The modelling power of variable i for class C is defined by

$$r_{i\{C\}} = 1 - \left(\frac{\sum_{n=1}^{N_C}(x_{in}-\hat{x}_{in\{C\}})^2/(N_C-P-1)}{\sum_{n=1}^{N_C}(x_{in}-\bar{x}_{i\{C\}})^2/(N_C-1)} \right)^{1/2} \qquad (7.27)$$

where $\bar{x}_{i\{C\}}$ is the mean of variable i over class C. This value ranges between 0 and 1, and the closer it is to 1, the more relevant it is for modelling the class. Another diagnostic statistic is the *discriminating power* of a variable between classes: the larger this value the better the discriminating power. In the example we discuss in this section neither of these diagnostics are very interesting since there are only two variables, but for multivariate datasets where many variables are measured these and related diagnostics can provide valuable aids and insight.

There are several software packages available for SIMCA and these usually provide the user with extensive diagnostics. This is possibly one reason for the popularity of SIMCA as a method of modelling classes in chemistry. In particular the original proponents of this approach, S. Wold and coworkers, published many of these theoretical developments in the chemical literature and applied the method first to analytical and organic chemical datasets. Most chemometricians know that it is possible to extend the list of statistical tests and diagnostics for most methods of data analysis. In commercial packages such as SAS there may typically be several hundred options or combinations of diagnostics for the most popular techniques, including more traditional statistical approaches such as discriminant analysis and cluster analysis, so SIMCA should not be chosen simply because of the large number of test statistics available in common implementations. The difficulty with many other methods for modelling is that their use is restricted to specialized statistical packages and no one has invested the time to code in all the various diagnostics to user-friendly micro based software, yet the average laboratory based chemist is unlikely to have the time to invest in learning and understanding sophisticated statistical packages. It is, therefore, probably best to use SIMCA as an initial approach and move to other methods if time and the importance of the problem allows it.

Before leaving the topic of soft modelling in chemistry it is worth noting that there are several other methods. These include UNEQ developed by Massart and Coomans, SPHERE developed by Strouf, and PRIMA developed by Veress.[1] All these model classes independently, so extra classes can be added without changing the models (unlike hard modelling), and objects may have partial membership of different classes. There is not room to compare the merits or otherwise of these different approaches in this text, although it is certainly very interesting, and the limited number of publications contrasting different methods for soft modelling in chemistry is probably a reflection of limitations in comprehensive user-friendly software rather than any other reason. Some chemometricians develop a method, report it in the literature and then move onto another method, whereas others spend several years exploiting and developing a small number of methods. Both contributions are important to the development of the subject.

7.6 FURTHER METHODS FOR MULTIVARIATE PATTERN RECOGNITION AND DATA DISPLAY

So far, in this chapter, we have concentrated on discussing methods for classifying groups of samples. Although much of the classical work on

[1]See bibliography.

chemometric pattern recognition has been concerned with this problem, multivariate approaches for data display involving simplifying and summarizing experimental data are also useful to the experimenter. PCA is one approach, involving reducing datasets to a few principal components. Graphs of scores and loadings are discussed elsewhere in this text but help provide qualitative information on the data. Often the main qualitative trends and patterns help interpret chemical processes.

There are many other ways of *dimensionality reduction* besides PCA. We summarize a few of these below.

Correspondence factor analysis (CFA) is a method of scaling data that was originally mainly developed by French statisticians.[1] It is useful in various situations.

(1) Where there are categorical data as well as or instead of continuous numerical data. The data may involve measurements of colour, hardness etc., where the variables are coded by an integer (e.g. red=1, blue=2) rather than a continually varying measurement.

(2) Where there is an interest in the contrast between variables. For example, we might measure the length and width of an object. In normal principal components analysis the size of the object will be of major importance. The value of the score of the first principal component will be most if both width and length are large, and least if they are smallest. CFA will be most useful if the ratio of width to length is of prime interest, in other words if we want to classify objects into fat and long objects. In chemical terms, CFA will be most useful when we are interested in ratios of abundances of compounds rather than absolute concentrations.

(3) Where objects and variables are interchangeable. This is less common in chemistry, where the objects are usually physical samples (e.g. a lump of rock, or an extract from a tissue) and the variables generally correspond to concentrations of compounds or related spectroscopic and chromatographic measurements.

Of course, CFA can be used for any dataset. In the case of numerical data, this method can be thought of as a way for scaling. The raw measurements are converted to a new scaled matrix. First the predicted value for each reading is estimated as follows.

$$\hat{x}_{ij} = \left(\sum_{k=1}^{I} x_{kj} \times \sum_{k=1}^{J} x_{ik}\right) / \left(\sum_{k=1}^{I} \sum_{l=1}^{J} x_{kl}\right) \tag{7.28}$$

This averages out the variability in the dataset. For example, in Table 7.16, the first column represents $31 / 91 = 0.341$ of the overall data. Therefore the

---

[1] It has been reported in other countries' literature under different names, but the French originated name has predominated.

predicted value of the data in the first column is always 0.341 of the sum of the entire column. The expected value $\hat{x}_{11}$ is $24 \times 0.341 = 8.18$. Of course, Eq. (7.28) treats the rows and columns symmetrically, so we could say that row 1 represents 0.264 of the overall variability and so $\hat{x}_{11} = 31 \times 0.264 = 8.18$: there is no difference, and the numerical answer we obtain is identical. This contrasts with most of the methods for scaling prior to PCA (Section 7.2) in which the variables and objects are treated differently. The expected values are given in Table 7.16. A new matrix is computed involving calculating

$$x'_{ij} \quad = \quad (x_{ij} - \hat{x}_{ij}) / \hat{x}_{ij} \tag{7.29}$$

Table 7.16 - Scaling prior to correspondence analysis

Raw data			Total (row)
7	4	13	24
8	3	10	21
10	5	20	35
6	2	3	11
Total (column) 31	14	46	

Predicted data (no variability)			
8.18	3.69	12.13	24
7.15	3.23	10.62	21
11.92	5.38	17.69	35
3.75	1.69	5.56	11
31	14	46	

Scaled matrix		
−0.14	0.08	0.07
0.12	−0.07	−0.06
−0.16	−0.07	0.13
0.60	0.18	−0.46

This is equivalent to calculating a parameter called the χ^2 statistic. If variables are completely correlated this value will be zero. For example, if a dataset consists only of two variables, which in turn correspond to two mass spectral peaks arising from the same compound, which are always observed in the same ratio, but with a different absolute intensity related to the concentration of the compound, the resultant matrix will consist of zeros. If the *ratio* of these peaks varies (which may indicate the presence of two different compounds in the mixture), the matrix will be non-zero.

Once this new matrix is calculated, it is treated in the same way as PCA, and components can be calculated in the normal way. The resultant dimensions need to be interpreted slightly differently, though, and will principally be concerned with variations in contrasts between samples and variables. Often CFA is a valuable complement to PCA for viewing major trends in large datasets. An important limitation of CFA is that it can only be used with positive data, since it is concerned with variability in proportions.

A related technique, devised by Lewi, is called *spectral map analysis*. The initial steps are as follows.

(1) The data are transformed to logarithms.[1]
(2) If there are negative logarithms, the most negative value is subtracted from all the data, to ensure that the overall matrix contains only positive elements.
(3) Each reading is divided by the overall matrix total so that each reading becomes a proportion of the entire matrix.
(4) The matrix is "double centred". This means that, both the row mean and the column mean are subtracted from each element in the data. These are the means after step 3 has been performed: the order in which these operations are performed is immaterial.
(5) The data are then treated in the same way as PCA or CFA.

A feature of spectral map analysis is that both contrasts and size are represented in the resultant diagram. For examples and further details, the reader is referred to the bibliography at the end of this chapter.

Another interesting method of data display is a *biplot* in which both scores and loadings are placed on the same graph. This may be useful in cases where there are several variables and roughly equivalent numbers of samples: relationships between the variables may often be similar to relationships between the samples. For example chemical constituents (variables) and geographical origins (samples) of river water might be equally diagnostic of whether the water is polluted: we want to know whether water is close to a source of pollution or if it contains high quantities of chemical pollutants – in some cases it makes sense

---

[1] If there are values close to zero, they are substituted by small positive numbers to avoid the problem of large negative logarithms.

to visualise both relationships at the same time. Biplots are often used by exponents of correspondence analysis, where variables and samples have the same meaning, but there is no necessity to represent the results of CFA by a biplot. Most chemometricians have steered away from biplots because variables and samples often have very different meanings to the chemist.

Non-linear mapping or *multidimensional scaling* is another form of dimensionality reduction that attempts to retain absolute distances between objects as much as possible. A stress function is computed that indicates the distortion in distance and is defined by

$$ S = [\sum_{n=2}^{n=N} \sum_{m=1}^{m=n-1} (d_{nm} - d_{nm}')^2] / \sum_{n=2}^{n=N} \sum_{m=1}^{m=n-1} (d_{nm})^2 \qquad (7.30) $$

where d_{nm} is the distance between objects n and m before dimensionality reduction[1] and d_{nm}' is the distance afterwards (normally in two or three dimensions corresponding to the two or three new variables). These variables are linear combinations of the original variables in exactly the same manner as PCA or CVA but computed using a different criterion.

Feature or variable selection is another area of interest. In Section 7.5, we discussed how the modelling or discriminating power of variables could be determined using SIMCA. This can help chemical interpretation. For example if we know that variable 7 corresponds to an alkene of known structure and is highly significant in determining the difference between two classes of samples, we may be able to interpret the relevant processes in chemical terms. If two classes correspond to two types of food, then this gives us a clue as to the chemical basis of taste, and can help determine what chemical additives are most suitable for use as additives in synthetic foodstuffs. Similarly it is possible to determine those chemicals or groups of chemicals that cause most variability, and, instead of using principal components, individual *markers* or groups of markers can be used instead. This approach has a psychological advantage in convincing non-chemometricians that multivariate methods for pattern recognition have a use in their own applications area. A good example is geochemistry, where scientists have for many years interpreted quite detailed physical and chemical processes using biomarkers, either a single compound or a ratio between two compounds extracted from sediments. Often very elaborate physical models are built from the change in these univariate parameters in geological time. To an extent this approach is highly empirical and is unlikely to accurately represent complex interacting multivariate natural processes, but it is based on simple chemical intuition. As a first stage, geochemists need educating away from these univariate parameters, but using abstract statistical components is often too much of a culture shock, so the use of multivariate methods to

---

[1]Normally the Euclidean distance (Eq. (7.5)).

determine which chemicals contribute most to variability is a first phase in educating applications scientists in these and related areas. Elaborate software has been produced in several applications areas, although it is questionable whether the motives are chemometric or the result of inevitable pressure to obtain grants from or publish in journals in the relevant applications area: there is not room in this text to review the interface between chemometrics and scientific computing in the various possible applications areas (archaeometry, chromatography, pharmacology, environmental chemistry and so on also contain rich possibilities). Probably the emphasis on individual marker compounds is a transition to truly multivariate models of complex natural processes, and, with increasing awareness of the motivations behind chemometrics, applications scientists will eventually adopt truly chemometric methods, instead of relying on univariate models for processes that are undoubtedly too complex to be interpreted in such a simplistic way.

This section has summarized a few of the many possible approaches to display and interpretation of multivariate patterns. The summary is undoubtedly subjective, and does not advocate that any method is better than any other method. The historical development of many computationally based fields, though, involves various front line investigators selling their software and methods as the answer to most experimental problems. Within chemical pattern recognition this trend is much in evidence, with various individuals and groups advocating their own methods. The approaches that are most cited in the literature and frequently reported in conferences are often those whose proponents concentrate most on marketing their ideas. There is no guarantee that these methods are best, but in the absence of user-friendly software, less experienced workers are probably best advised to stick to a few well established methods for which there is properly documented, commercially available, and professionally supported software. It is worth looking at software reviews in chemometrics and related journals and choosing packages that are marketed by large firms that are likely to be in business for many years. Telephone support in case of bugs and difficulties in the software is also important and can save a great deal of time and effort in the long run. Relatively new methods are principally of interest to a small group of practicing chemometricians and it takes several years for approaches to become accepted and developed in the form of commercially available user-friendly software.

7.7 FUZZY METHODS

A very different approach to pattern recognition comes from *fuzzy set theory*. Fuzzy methods can, of course, be used in all areas of chemometrics but are described in the context of pattern recognition because the best known chemical applications are in this area.

In most of the methods discussed so far in this chapter, measurements, often in the form of spectroscopic or chromatographic peak intensities, are assumed to be recorded at discrete intervals. For example, an infrared spectrum may be recorded at wavelengths 1500, 1510, 1520 cm^{-1} and these intensities are then compared to intensities of pure compounds or classes of compounds and so enable the spectroscopist to make deductions about the chemical origin of the spectrum. Because of noise and blurring, spectra, even of pure compounds, are not completely reproducible. One major problem is that linewidths can sometimes be dependent on recording conditions rather than chemistry. This process of blurring has been discussed in Chapter 5 in the context of univariate signal processing and the maximum entropy approach is discussed in Section 5.4. Fuzzy methods are another way of dealing with this problem. A second problem is that the position of peaks is not always reproducible. This can be due to chemical factors but is often due to problems of exact calibration of wavelengths or time, dependent on the type of spectroscopy. In chromatography, of course, this problem is severe and retention times are not exactly reproducible.

So, if the training set for a series of compounds leads us to expect a maximum at 1510 cm^{-1} but, because of instrumental errors, the maximum in an unknown spectrum is actually 1506 cm^{-1}, if we use the wavelength at 1510 cm^{-1} as input to a pattern recognition program, the intensity we use will be slightly less than the true intensity. In most pattern recognition programs this variation of intensity is assumed to be caused by noise, and the shift in maximum or width of peaks is not taken into account. Although signal processing and deconvolution methods can improve the quality of data available to pattern recognition software, another approach is to model the changes in shape and positions of peaks rather more empirically, using fuzzy methods, as discussed below.

Instead of assuming the centre of the peak must occur at exactly 1510 cm^{-1} an alternative approach is to assume that the peak has a probability of being centred at 1510 cm^{-1} and slightly lower probabilities of being centred at 1509 and 1511 cm^{-1}, and so on. This probability distribution is called a *membership function* (m.f.). We can mathematically define the m.f. however we like. A simple example is a triangular function, but more elaborate models such as

Gaussians and Lorentzians (Cauchy distributions)[1] can be used. Fig. 7.18 represents a predicted chromatogram or spectrum consisting of five peaks (Table 7.17) in which each peak is indicated by a sharp spike. A triangular membership function in which

$$m_i(x) \;=\; [\,(1-|x-c_i|/2) + |1-|x-c_i|/2|\,]/2 \qquad (7.31)$$

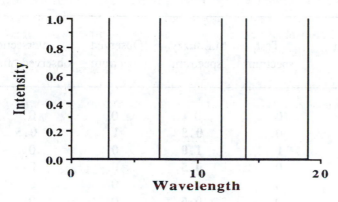

Fig. 7.18 - Predicted (crisp) spectrum (Table 7.17).

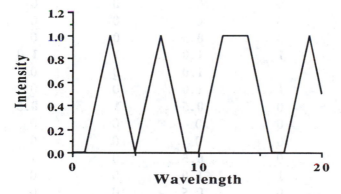

Fig. 7.19 - Membership function for spectrum in Fig. 7.18.

can be computed for each peak, i, centred at $x = c_i$. The membership functions are illustrated in Fig. 7.19, and calculated in Table 7.17. An unknown spectrum, consisting of four peaks, is illustrated in Fig. 7.20. Is this consistent with the spectrum of Fig. 7.18? The way to test this is by calculating the intersection between the observed spectrum (usually called the *crisp* spectrum in the fuzzy literature) in Fig. 7.20 and the fuzzy spectrum of Fig. 7.19, defined by

[1]See Section 5.6 for definition.

$$m_{i \cap j}(x) \; = \; \min \, (m_i(x), m_j(x)) \tag{7.32}$$

This is illustrated in Fig. 7.21 and the calculation given in Table 7.17. The union between two sets[1] is defined as

$$m_{i \cup j}(x) \; = \; \max \, (m_i(x), m_j(x)) \tag{7.33}$$

Table 7.17 - Example of fuzzy set calculation

Wavelength	Test spectrum	M.f. fuzzy spectrum	Observed spectrum	Intersection observed∩fuzzy
1	0	0	0	0
2	0	0.5	1	0.5
3	1	1.0	0	0
4	0	0.5	0	0
5	0	0	0	0
6	0	0.5	0	0
7	1	1.0	1	1.0
8	0	0.5	0	0
9	0	0	0	0
10	0	0	0	0
11	0	0.5	0	0
12	1	1.0	1	1.0
13	0	1.0	0	0
14	1	1.0	0	0
15	0	0.5	1	0.5
16	0	0	0	0
17	0	0	0	0
18	0	0.5	0	0
19	1	1.0	0	0
20	0	0.5	0	0

Finally we introduce the concept of *cardinality*,

$$\text{card}_i \; = \; \Sigma \, m_i(x) \tag{7.34}$$

which is merely the sum or integral (for continuous distributions) of the membership function i over x. Sometimes there are limits to the summation. In the language of fuzzy sets, x is part of a Universe X (this Universe may be of

---

[1]Not used in the present calculations but employed in other operations in the fuzzy literature.

frequencies, time, wavelength, concentration etc.) and the cardinality is calculated over this Universe. *Relative cardinality* is defined as

$$\text{rel}_U \text{card}_i \quad = \quad \text{card}_i / \text{card}_U \qquad\qquad (7.34)$$

where U is a standard set, normally the Universe X. For example, the relative cardinality of a function that can take any value between 0 and 1 between wavelengths 1500 and 1599 cm^{-1} will be the overall cardinality divided by 100 (maximum possible cardinality). The maximum possible cardinality for the spectrum of Table 7.17 is 5, since there are five peaks of intensity 1 in the predicted chromatogram. The intersection of the fuzzy spectrum and the observed spectrum has a cardinality of 3 (two peaks of intensity 1 and two of intensity 0.5) (Fig. 7.21), so the relative cardinality is 0.6.

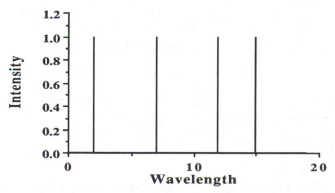

Fig. 7.20 - Unknown (observed) spectrum to be tested for compatibility
with the spectrum of Fig. 7.18.

This approach can be used in more complex situations, to classify spectra or chromatograms. For example, each class might be represented by a fuzzy spectrum or chromatogram, representing typical peaks expected to occur in the class. Observed or test spectra can then be tested against the fuzzy spectrum, and the relative cardinality of the intersection used to classify the unknown. The higher the cardinality, the more likely it is to belong to a particular class. Thus relative cardinality may be used as a class membership function.

It is possible to define membership functions in more than one dimension. The second dimension could be intensity, and the first position, for example. A tetrahedral m.f. may be defined; alternatively a two dimensional Gaussian shape may be used. It is not always desirable to fuzzify the known (training or candidate) spectrum. The unknown spectrum could be fuzzified instead, and the intersection with several crisp candidate spectra tested. This latter approach might be useful for searching spectroscopic libraries where a best match between several hundred candidate spectra is desired. The three or four spectra with the highest relative cardinality could then be presented to the user who would manually choose the best answer.

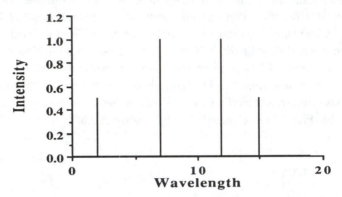

Fig. 7.21 - Fuzzy intersection of Fig. 7.19 and 7.20.

Fuzzy methods can also be used in cluster analysis where each object is assigned a membership function for a cluster. This is superficially similar to SIMCA where each object is assigned a distance to a group or a cluster. However, there exists much less software for fuzzy clustering compared to soft modelling in chemistry.

Fuzzy methods are by no means restricted to pattern recognition. Applications as diverse as mixture analysis, calibration and modelling have been studied using fuzzy set theory. In some cases fuzzy methods give a correct answer more frequently than conventional approaches, but a great deal depends on the nature of the membership function. Obviously elaborate m.f.s can be constructed with a good knowledge of noise distributions, but clearly if everything is known about the noise distribution a great deal is already known about the system and other approaches might be more successful. It is fairly difficult assessing the effectiveness of fuzzy methods on simulated or model datasets since it is always possible, and tempting, to improve the membership function to get better answers. Simple changes, such as of the width of Gaussian or triangular membership functions will make a difference to the overall answer. In real situations we have only partial knowledge of the system and may not have the time or information to produce a detailed model for the membership function.

One area in which fuzzy methods almost certainly have an advantage over conventional methods is in linear calibration in the presence of outliers, i.e. readings that probably do not correspond to the underlying model and so are very far from the straight line. A best fit straight line may be defined as the straight line whose intersection with the fuzzy membership function of the raw datapoints is a maximum. Outliers will have a low membership function so the straight line will be more sensitive to points lying close to the straight line. In conventional linear calibration using least squares error criteria, the outliers will dominate the analysis.

Although fuzzy methods have a strong potential use, they should be contrasted with maximum entropy methods where a dominant theme is deblurring rather than blurring spectra or images. Maximum entropy has been used far more widely in chemical data analysis: this is largely because of the availability of good software, and fuzzy methods will remain an interesting curiosity until far more resources are invested in the development, validation and marketing of robust user-friendly software.

7.8 TIME SERIES ANALYSIS

Finally, we will discuss the topic of naturally occurring time series. In various sections of this text we have discussed continuously varying cyclical processes (Sections 3.4, 4.7 and 5.2). Fourier transforms were introduced as a method for analysing these time series. In the case of spectroscopy and most other instrumental measurements, Fourier transformation of the raw time series together with phasing and possibly some form of filtering, as discussed in Chapter 5, is sufficient. However, time series arise under several different circumstances in chemometrics. Cyclical events occur widely in nature: long term geochemical processes, environmental processes dependent on diurnal, annual, monthly or tidal rhythms and clinical processes dependent on regular hormonal rhythms are all good examples. The chemist may analyse these events by extracting samples taken at regular intervals and using spectroscopic or chromatographic techniques for estimating the concentration of compounds in these samples. Sometimes several compounds can be measured in each sample, resulting in a truly multivariate time series.

Straight Fourier transformation of such time series often results in a very noisy spectrum. The problem is that the noise is rarely white non-stationary but is often coloured, that is correlated in time, often modelled as an ARMA process, as discussed in Section 3.2. This is fairly likely in many natural situations. Consider the case of long term cyclical changes in temperature over geological time. There are various theories that suggest that one major factor influencing these changes is the changing orientation of the orbit of the earth around the sun, which can be modelled by a sum of harmonics of various frequencies. The difficulty here is that there is bound to be some other random factor that also influences the temperature. If this elevates the temperature by 5° over the (cyclical) predicted temperature at a given time, at the next sampling interval we expect the temperature to be $(5+\delta)^{o}$ above the predicted temperature, where δ is a random number (that can take negative as well as positive values), so the absolute value of the error at each sampling interval is partly dependent on the absolute value at the previous sampling interval. This partially correlated component of the time series is called a *non-deterministic* component. The cyclical component (which is the part of the time series of principal interest) can be exactly predicted at any time in the future and is called a *deterministic* component. According to the work of H. Wold every time series can be decomposed into a deterministic and non-deterministic part. The most common method for analysing such data is called *spectral analysis*: at this point it is important to realize that chemists and statisticians frequently use similar terminology to refer to different procedures. In strict statistical terms spectral

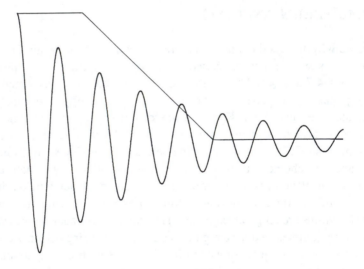

Fig. 7.22 - Simple time series filter.

analysis (a form of time series analysis) can be applied to any sequentially varying process, including long term voting trends, the price of a product over time, signals from electrical circuits and so on: it refers to a form of data analysis. The main steps in the procedure are as follows.

(1) Calculate a correlogram. For a single time series we compute an autocorrelogram as described in Section 3.2.

(2) Optionally filter the correlogram using one of several windows. The principle of filters is discussed in Section 5.6. There is a large literature on choice of

windows, to which the reader is referred in the bibliography. However, the choice of window depends, to a certain extent, on the number of datapoints in the correlogram or the maximum *lag*. If a raw time series consists of 100 datapoints, then if a correlogram is calculated from lag values of between 0 and 59, the 60th datapoint of the time series will involve only 40 of the original datapoints. The confidence in the correlogram decreases at increasing lag values, as less of the original information is used to compute the later datapoints. Simple filters involve multiplying by functions that decrease in time (Fig. 7.22) reflecting decreasing confidence in the data along the correlogram. Some care must be taken when applying "optimal" windows in the literature. The formulae are usually derived for infinitely long time series with a normal error distribution, and so are only approximations for short time series with coloured noise.

(3) Fourier transform the correlogram.

(4) Phase the resultant frequency domain spectrum or, more commonly, use the absolute value or power spectrum (see Section 5.2).

(5) Optionally use a smoothing function such as a moving average (Section 5.3) to improve the appearance of the resultant spectrum.

Common trends in two time series can be compared using an *cross-correlogram*. Instead of computing the autocorrelation coefficient (Eq. (3.4)) we calculate the cross correlation coefficient between variables x_1 and x_2, defined by

$$\rho_i = \frac{\displaystyle\sum_{t=1}^{N_2-i}[(x_{1,t}-\bar{x}_{1(N_2-i,1)}) \times (x_{2,t+i}-\bar{x}_{2(N_2,i+1)})]}{\sqrt{\displaystyle\sum_{t=1}^{N_2-i}(x_t-\bar{x}_{1(N_2-i,1)})^2 \displaystyle\sum_{t=i+1}^{N_2}(x_t-\bar{x}_{2(N_2,i+1)})^2}} \qquad (7.35)$$

Some explanation of the terms and meaning of this equation are required. It is assumed that $N_2 \leq N_1$ where N_1 is the number of observations of x_1 and N_2 of x_2. The term $\bar{x}_{k(n,m)}$ refers to the mean of x_k between points n and m. There is no requirement that the two time series contain the same number of terms, only that they are sampled equally regularly. The overlap between the two time series consists of N_2-i datapoints. If the first time series consists of 30 points (numbered 1 to 30), and the second of 20 points, then, if $i=5$, points 1 to 15 of x_1 and points 6 to 20 of x_2 are used in the calculation of ρ_i. This overlap is illustrated in Fig. 7.23. Unlike the autocorrelation coefficient, the standard deviations only of the overlapping parts of the chains are used in the calculation of cross-correlation coefficients. The cross-correlation coefficient is not symmetrical: Eq. (7.35) refers only to positive values of i. For $i < 0$, it is necessary to exchange x_1 and x_2 in Eq. (7.35).

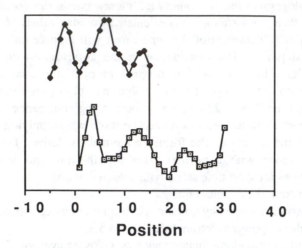

Fig. 7.23 - Example of two time series consisting of a different number of datapoints and a relative lag of 5. The area of overlap is indicated.

The size or "power" of the correlation coefficient in the resultant transform of the correlogram can be interpreted in terms of strength of cyclicity. The greater this number, the more likely the cyclicity; it is possible to convert to a probability scale.

Another major limitation of many naturally occurring time series is that it is difficult to sample regularly in time. Consider the problem of sampling a river every quarter of an hour. It may take 1 or 2 minutes to obtain each sample, so the size of the lot is significant in relation to the sampling frequency, introducing errors as discussed in Section 3.4. In addition technicians and apparatus may not always be exactly ready at regular intervals. People take lunchbreaks, and disasters happen over protracted periods of time, meaning that the occasional irregular sample must be tolerated. Refusing to analyse a series of 40 samples because two of the samples happen to have been taken irregularly involves a massive loss of information. It is often necessary to *interpolate* from an irregularly sampled set of samples to estimate a regularly sampled time series, since it is only possible to compute correlograms and perform Fourier analysis on regularly recorded data. There are several methods for interpolation, but the simplest one is linear interpolation. If we want to estimate the value of x at time t $(= \hat{x}_t)$ then if the sample immediately prior to time t was taken at $t-\delta_1$ and the sample immediately afterwards at $t+\delta_2$,

$$\hat{x}_t \quad = \quad (\delta_1 x_{t+\delta_2} + \delta_2 x_{t-\delta_1})/(\delta_2+\delta_1) \tag{7.36}$$

More elaborate interpolation functions may be proposed, including those that are jointly used for estimation and smoothing. Some such approaches are discussed in Chapter 3, but some care must be taken if sampling close to the Nyquist frequency. For polynomials it does not matter too much if the time series is smoothed away, but in the case of rapidly oscillating sine waves oversmoothing can have disastrous effects and result in complete loss of meaningful information at high frequency.

A further problem involves interpreting time series when samples are not directly recorded as a function of time. This is common in geochemical situations where cores are sampled in depth. It then is necessary to calibrate depth to time, since the natural processes are cyclically related to time rather than depth. This can be a difficult problem as depth is unlikely to be a linear function of time. There are several procedures and tremendous debate in the literature as to the best approach. One method involves establishing known dates and linearly interpolating the time scale between these events. Other, more elaborate, approaches involve optimizing the time scale to maximize cyclicity in the resultant spectrum. The problem with these methods for tuning is that they already assume a model and it is questionable how useful the results are. Simpler approaches may not produce quite as good spectra but are probably more robust.

Finally, there are often non-cyclical components in naturally occurring time series. These can be removed by *detrending*. A straight line or polynomial is fitted to the time series before computing the correlogram and is subtracted from the data. The residuals are then used for subsequent spectral analysis.

There is no set order in which the steps in spectral analysis of naturally occurring time series take place, but a typical order[1] is as follows:
(1) Detrend the time series.
(2) Convert from a depth to a time scale.
(3) Interpolate to a regular sampling interval in time.
(4) Multiply by a window function.
(5) Compute the correlogram.
(6) Multiply by a Fourier filter function.
(7) Fourier transform.
(8) Phase or calculate absolute value and / or power spectrum.
(9) Smooth the resultant spectrum.

It is necessary to interpret probabilities from the resultant spectrum with some caution because of the numerous stages in the analysis. Another problem is that most probabilities cited are dependent on various models of noise which are quite likely to be inappropriate under these circumstances.

---

[1]Some steps are omitted in some applications.

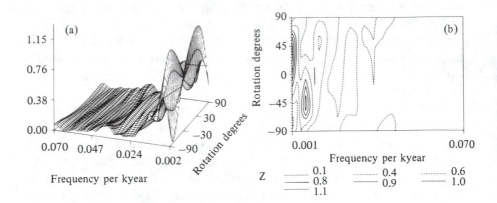

Fig. 7.24 - Spectral analysis on a multivariate time series. The spectra of the rotated principal components are calculated.

Another possibility is the analysis of multivariate time series, where several compounds are recorded at each time interval. Instead of performing spectral analysis on the concentration of each individual compound, it is possible to compute principal components and perform spectral analysis on these components. There is no general guideline as to what is the best stage in the analysis to compute the principal components, providing it is done before Fourier transformation. It is also possible to rotate principal components, in exactly the same way as in factor analysis (Section 6.3), and to transform these rotated components. Fig. 7.24 is an example of the correlogram of the concentrations of chemicals with time in a geological core, after several of the data analytical steps discussed above, in which spectral analysis is performed on the rotated principal components

$$P_r \quad = \quad \cos\theta \, P_1 + \sin\theta \, P_2 \qquad\qquad (7.37)$$

where P_1 is the first principal component and P_2 is the second principal component.

The resultant transform appears to separate the spectrum into components of different frequencies, roughly at right angles to each other.

There is a great deal of theory of multivariate time series analysis, much of it due to Box and Jenkins, which remains to be applied in chemometrics. This particular area promises interesting future possible developments in chemical pattern recognition.

REFERENCES

General
P.C.Jurs and T.L.Isenhour, *Chemical Applications of Pattern Recognition*, Wiley, New York, 1975
K.Varmuza, *Pattern Recognition in Chemistry*, Springer-Verlag, Berlin, 1980
O.Strouf, *Chemical Pattern Recognition*, Research Studies Press, Letchworth, 1986
R.A.Johnson and D.W.Wichern, *Applied Multivariate Statistical Analysis, Second Edition*, Prentice-Hall, Englewood Cliffs, New Jersey, 1988
C.Chatfield and A.J.Collins, *Introduction to Multivariate Analysis*, Chapman and Hall, London, 1989
B.Flury and H.Riedwyl, *Multivariate Statistics: a Practical Approach*, Chapman and Hall, London, 1988
B.F.J.Manly, *Multivariate Statistical Methods: a Primer*, Chapman and Hall, London, 1986

Section 7.2
O.M.Kvalheim, *Anal. Chim. Acta*, **177**, 71 (1985)
J.Aitchison, *The Statistical Analysis of Compositional Data*, Chapman and Hall, London, 1986

Section 7.3
D.L.Massart and L.Kaufman, *The Interpretation of Analytical Chemical Data by the Use of Cluster Analysis*, Wiley, New York, 1983
N.Bratchell, *Chemometrics Int. Lab. Systems*, **6**, 105 (1989)
B.S.Everitt, *Cluster Analysis*, Heinemann, London, 1974
C.T.Zahn, *IEEE Trans. Comput*, **20**, 68 (1971)

Section 7.4
N.J.Nilsson, *Linear Learning Machines*, McGraw-Hill, New York, 1965
B.R.Kowalski and C.F.Bender, *J. Am. Chem. Soc.*, **94**, 5632 (1972)
R.A.Fisher, *Annals of Eugenics*, **7**, 179 (1936)
D.Coomans, M.Jonckheer, D.L.Massart, I.Broeckaert, P.Blockx, *Anal. Chim. Acta*, **103**, 409 (1978)

Section 7.5
S.Wold, *Pattern Recognition*, **8**, 127 (1976)
S.Wold and M.Sjøstrøm, in B.R.Kowalski (Editor), *Chemometrics: Theory and Practice*, ACS Symposium Series 52, p. 243 (1977)
O.Strouf and J.Fusek, *Collect. Czech. Chem. Commun.*, **44**, 1370 (1979)
M.P.Derde and D.L.Massart, *Anal. Chim. Acta*, **184**, 33 (1986)
V.Juricskay and G.E.Veress, *Anal. Chim. Acta*, **171**, 61 (1985)

Section 7.6
P.J.Lewi, *Chemometrics Int. Lab. Systems*, **5**, 105 (1989)
M.Greenacre, *Theory and Application of Correspondence Analysis*, Academic Press, London, 1984
S.S.Schiffman, M.L.Reynolds and F.W.Young, *Introduction to Multidimensional Scaling*, Academic Press, Orlando, Florida, 1981

Section 7.7
M.Otto, *Chemometrics Int. Lab. Systems*, **4**, 101 (1988)

Section 7.8
R.G.Brereton, *Chemometrics Int. Lab.Systems*, **2**, 177 (1987)
C.Chatfield, *The Analysis of Time Series: An Introduction*, Chapman and Hall, London, 1984
G.M.Jenkins and D.G.Watts, *Spectral Analysis and its Applications*, Holden-Day, San Francisco, California, 1968
G.E.P.Box and G.M.Jenkins, *Time Series Analysis, Forecasting and Control*, Holden-Day, San Francisco, California, 1970

Appendix

A.1 INTRODUCTION

This appendix includes some brief mathematical details and also computational background of various formulae and algorithms in various chapters of this text. There are many excellent mathematical texts on statistics to which the reader is referred for further details.

A.2 FUNCTIONS OF SCALARS

In this text we denote the following functions of x.
(1) The *mean* is represented as $\bar{x}$. Unless otherwise stated, this mean is over the entire dataset.
(2) The *estimated* value is represented by $\hat{x}$.
(3) Transformations of x are denoted by x' and x'' where appropriate.

Sums of squares are denoted as follows, unless otherwise stated.
(1) The sum of squares for a dataset is S.
(2) The variance (or mean square) is s^2.
(3) The root mean square is s.

A.3 VECTORS AND MATRICES

A vector is a one dimensional array of numbers. A *column* vector is of the form

$$\mathbf{x} \quad = \quad \begin{pmatrix} 1 \\ 3 \\ 4 \end{pmatrix} \qquad\qquad (A.1)$$

The numbers 1, 3 and 4 may represent the intensity of a *variable* (e.g. the height of a spectroscopic peak) measured on three samples. A *row* vector is of the form

$$\mathbf{y} \quad = \quad (1\ 7\ 9\ 10) \qquad\qquad (A.2)$$

and might represent four measurements on one sample.

A *matrix* is two dimensional and is of the form

$$\mathbf{X} \quad = \quad \begin{pmatrix} 1 & 7 & 9 & 10 \\ 3 & 5 & 4 & 3 \\ 4 & 12 & 6 & 5 \end{pmatrix} \tag{A.3}$$

This matrix is a 3×4 matrix: the first dimension quoted is the the number of rows.

Generally matrices are denoted by upper case bold characters and vectors by lower case bold characters. The elements of the matrices are called *scalars* and are single numbers. Sometimes they are indicated by subscripts so that x_{23} may be used to denote the element of the second row and third column, which equals 4 in the example above.

One historical feature of chemical dataprocessing is that objects are normally represented by rows and variables by columns. This is in contrast to many other application areas. In this text, to avoid confusion, we retain this convention, but this means that there is some inconsistency in notation. If the reader wants to convert the equations in this text to matrix notation he or she must carefully check the meaning of the subscipts of scalar variables. Generally variables are denoted by the letters i, j etc. and experiments, objects or samples, by the letters n, m etc. Because we commonly refer to x_i as denoting all the measurements of variable i over an entire dataset, the value of this variable for experiment n is called x_{in} rather than x_{ni}. Since matrices are rarely used in this text there should be no confusion with these conventions. More mathematically rigorous conventions would lead to loss of readability.

The *transpose* of a matrix is the matrix that is formed from interchanging the rows and columns, and in this text we will denote the transpose of $\mathbf{X}$ by $\mathbf{X'}$ so that

$$\mathbf{X'} \quad = \quad \begin{pmatrix} 1 & 3 & 4 \\ 7 & 5 & 12 \\ 9 & 4 & 6 \\ 10 & 3 & 5 \end{pmatrix} \tag{A.4}$$

in the example above.

Matrix addition or subtraction is easy, but each matrix must have identical number of rows and columns. The elements of the new matrix equal the sum (or difference) of the elements of the constituent matrices, so, for example

$$\begin{pmatrix} 5 & 3 \\ 2 & 8 \\ -3 & 1 \end{pmatrix} + \begin{pmatrix} 1 & -3 \\ 6 & 0 \\ -2 & 5 \end{pmatrix} = \begin{pmatrix} 6 & 0 \\ 8 & 8 \\ -5 & 6 \end{pmatrix} \tag{A.5}$$

Matrix multiplication is slightly more complicated. The order in which two matrices are multiplied is important so

$$\mathbf{A} \cdot \mathbf{B} \neq \mathbf{B} \cdot \mathbf{A} \tag{A.6}$$

The number of columns in the first matrix must equal the number of rows in the second matrix; the number of rows in the product matrix equals the number of rows in the first matrix and the number of columns in the product matrix equals the number of columns in the second matrix. Therefore if $\mathbf{A}$ is a 3×2 matrix and $\mathbf{B}$ is a 2×4 matrix, the product

$$\mathbf{A} \cdot \mathbf{B} = \mathbf{C} \tag{A.7}$$

is a 3×4 matrix.

The element c_{nm} of the product is given by

$$c_{nm} = \sum_{i=1}^{i=2} \sum_{j=1}^{j=2} a_{ni} b_{jm} \tag{A.8}$$

in the case above, so that

$$\begin{pmatrix} 5 & 8 \\ 3 & 1 \\ -2 & -4 \end{pmatrix} \cdot \begin{pmatrix} 1 & 0 & -2 & 4 \\ 3 & 2 & 5 & -6 \end{pmatrix} = \begin{pmatrix} 29 & 16 & 30 & -28 \\ 6 & 2 & -1 & 6 \\ -14 & -8 & -16 & 16 \end{pmatrix} \tag{A.9}$$

A *square matrix* is one in which the number of rows equals the number of columns. A special matrix is the *unit matrix* in which the diagonal elements all equal exactly 1 and all other elements equal 0. We will denote this matrix by $\mathbf{I}$. The *inverse* of a square matrix $\mathbf{A}$ is denoted by $\mathbf{A}^{-1}$ and defined by

$$\mathbf{A}^{-1} \cdot \mathbf{A} = \mathbf{A} \cdot \mathbf{A}^{-1} = \mathbf{I} \tag{A.10}$$

There are various computational methods for calculating inverses which need not worry us here. It is easy to verify that the inverse of

$$\mathbf{A} = \begin{pmatrix} 1 & -3 & 6 \\ 5 & 4 & -5 \\ 7 & 0 & -1 \end{pmatrix} \tag{A.11}$$

is given by

$$\mathbf{A}^{-1} = \begin{pmatrix} 0.049 & 0.037 & 0.110 \\ 0.366 & 0.524 & -0.427 \\ 0.341 & 0.256 & -0.232 \end{pmatrix} \tag{A.12}$$

Some matrices do not have inverses and are called non-singular. There are various ways of getting around this problem, including using what is called the Moore-Penrose generalized inverse.

Inverse matrices are only defined for square matrices. The operation of matrix division is not defined.

A.4 PRINCIPAL COMPONENTS ANALYSIS: TWO VARIABLES

As discussion in Section 4.2, PCA for two variables differs from standard linear regression of one variable against another one in that errors in both variables are assumed significant. Therefore, standard linear regression software will not provide a PCA solution. It is possible to deduce an algebraic formula for the PCA solution as follows.

We assume that the two new variables are

$$x_1' = (\cos \theta)\, x_1 + (\sin \theta)\, x_2 \tag{A.13}$$

$$x_2' = -(\sin \theta)\, x_1 + (\cos \theta)\, x_2 \tag{A.14}$$

We want to minimize the sum of squares

$$S = \sum_{n=1}^{n=N} x_2'^2 \tag{A.15}$$

where N is the number of samples in the dataset. Therefore we want to minimize

$$S = \sin^2\theta \sum_{n=1}^{n=N} x_1^2 - 2\sin\theta\cos\theta \sum_{n=1}^{n=N} x_1 x_2 + \cos^2\theta \sum_{n=1}^{n=N} x_2^2 \tag{A.16}$$

So we want a solution to the equation $\partial S / \partial\theta = 0$ or

$$2\sin\theta\cos\theta \left[\sum_{n=1}^{n=N} x_1^2 - \sum_{n=1}^{n=N} x_2^2 \right] - 2(\cos^2\theta - \sin^2\theta) \sum_{n=1}^{n=N} x_1 x_2 = 0 \tag{A.17}$$

which gives (using standard trigonometric identities)

$$\tan 2\theta = \frac{2 \sum_{n=1}^{n=N} x_1 x_2}{\left[\sum_{n=1}^{n=N} x_1^2 - \sum_{n=1}^{n=N} x_2^2 \right]} \tag{A.18}$$

where θ is the angle of rotation for the second principal component. The first principal component is merely at right angles to this one.

A.5 LINEAR DISCRIMINANT ANALYSIS: TWO CLASSES

An example of linear discriminant analysis or canonical variates was discussed in Section 7.4. In this section we show how to calculate the function for two classes, which should maximize the function in Eq. (7.12).

We assume that there are N_A observations in class A and N_B observations in class B, and that I variables have been measured. The vector of means for class A is given by

$$\bar{\mathbf{x}}_A = (1/N_A) \sum_{n=1}^{n=N_A} \mathbf{x}_n \qquad (A.19)$$

where the column vector $\mathbf{x}_n$ is defined.

$$\mathbf{x}_n = (x_{1n}, x_{2n}, \ldots x_{In}) \qquad (A.20)$$

i.e. is the vector of I variables measured for observation n, and observations in class A are numbered from 1 to N_A. The variance-covariance matrix for class A is defined by

$$\mathbf{S}_A = (1/[N_A-1]) \sum_{n=1}^{n=N_A} (\mathbf{x}_n - \bar{\mathbf{x}}_A)' (\mathbf{x}_n - \bar{\mathbf{x}}_A) \qquad (A.21)$$

Similar definitions apply to class B.

The pooled variance-covariance matrix between the two classes is given by

$$\mathbf{S}_{AB} = \frac{(N_A-1)\,\mathbf{S}_A + (N_A-1)\,\mathbf{S}_B}{(N_A-N_B-2)} \qquad (A.22)$$

The linear discriminant function is defined by

$$x_1' = (\bar{\mathbf{x}}_A - \bar{\mathbf{x}}_B)\,\mathbf{S}_{AB}^{-1}\,\mathbf{x}' \qquad (A.23)$$

where $\mathbf{x}$ is the original vector of measurements. For the case of Table 7.11, consisting of two variables, the discriminant function is calculated in Eq. (7.13).

More complex mathematical algorithms to calculate two or more canonical variates are outside the scope of this text.

A.6   PRINCIPAL   COMPONENTS   ANALYSIS: NIPALS ALGORITHM

There are various methods for calculating the principal components of datasets consisting of more than two variables. Most approaches calculate several components simultaneously, but in cases where memory is limited and slow (e.g. microcomputers) iterative methods that compute one component at a time have been developed. It is possible to assess the significance of each component using methods such as cross-validation as described in Section 6.2. We describe the NIPALS algorithm, originally developed by S. Wold, below.

The matrix of samples (rows) and variables (columns), $\mathbf{X}$, is column centred, so that the mean of each variable (or column) is 0. We will call this matrix $\mathbf{X_0}$. The column vectors, initially consisting of the values of variable i over N experiments, will be denoted by $\mathbf{x}_{i(0)}$. The row vectors, initially consisting of the values of I variables for experiment n will be denoted by $\mathbf{r}_{n(0)}$.

We start with matrix X_t where t refers to the current residual matrix and $t=0$ initially. The following steps are performed.

(1) Select the column (factor) of X_t with the greatest sum of squares: this is a first guess of the principal component scores. Call this vector $\hat{x}_{max(t+1)}$ and call the sum of squares of its elements $\hat{x}^2_{max(t+1)}$

(2) Calculate the row vector

$$\hat{r}_{max(t+1)} = \hat{x}_{max(t+1)}' X_t / \hat{x}^2_{max(t+1)} \qquad (A.24)$$

and call the sum of squares of the elements of this vector $\hat{r}^2_{max(t+1)}$

(3) Scale the new vector to unit length

$$\hat{r}_{max(t+1)} = \hat{r}_{max(t+1)} / \hat{r}_{max(t+1)} \qquad (A.25)$$

(4) Calculate a new estimate of the principal component scores

$$\hat{x}_{max(t+1)} = X_t \hat{r}_{max(t+1)}' / \hat{r}^2_{max(t+1)} \qquad (A.26)$$

(5) Re-estimate the sum of squares for the new estimate of the principal component, i.e. calculate

$$S = \hat{x}^2_{max(t+1)} \qquad (A.27)$$

(6) Compare this sum of squares to that of the previous estimate. If the difference between these two estimates is small (perhaps $0.001\ S$) then the algorithm has converged for this principal component, and we go to step 7. Otherwise return to step 2 until convergence. When converged $\hat{x}_{max(t+1)}$ equals the scores and $\hat{r}_{max(t+1)}$ equals the loadings for component $t+1$.

(7) Calculate the new value

$$X_{t+1} = X_t - \hat{x}_{max(t+1)} \hat{r}_{max(t+1)}$$ and then repeat the algorithm, increasing t by 1, as from step 2 in order to obtain further principal components.

It is easier to compute the sum of squares for a vector by multiplying the vector by its transpose, but, of course these expressions could be expressed as sums of squares. It is usual to describe computational algorithms in terms of matrix operations.

A.7 PARTIAL LEAST SQUARES ALGORITHM: ONE Y VARIABLE

There are two main algorithms for PLS, one, due to S. Wold, non-iterative providing there is only one y variable (PLS1), and the other due to H. Martens which is iterative. For further details, the reader is referred to the bibliography of Section 6.4.

The matrix of measurements is first column centred to give a new matrix which we will call X_0; the y vector, consisting of N measurements (e.g. concentrations) for each of the samples is also column centred to give a new vector y_0.

We will start with the the matrix and vector X_t and y_t where $t = 0$ initially. The following steps are performed.

(1) Calculate the loading weights
$$w = X_t' \, y_t / \sqrt{(\, y_t' \, X_t \, X_t' \, y_t \,)} \qquad\qquad (A.28)$$

(2) Estimate the score for the next PLS component by
$$\hat{x}_{t+1} = X_t \, w \qquad\qquad (A.29)$$
Call the sum of squares of the elements of this new vector $\hat{x}_{t+1}^2$

(3) Estimate the loading for the next PLS component by
$$\hat{r}_{t+1}' = X_t' \, \hat{x}_{t+1} / \hat{x}_{t+1}^2 \qquad\qquad (A.30)$$

(4) Estimate the value of the residual contribution to y by
$$\delta y_{t+1} = \hat{x}_{t+1} \, (y_t' \, \hat{x}_{t+1}) / \hat{x}_{t+1}^2 \qquad\qquad (A.31)$$

(5) Calculate the new values
$$X_{t+1} = X_t - \hat{x}_{t+1} \, \hat{r}_{t+1} \text{ and } y_{t+1} = y_t - \delta y_{t+1}$$ and repeat the algorithm, increasing t by 1, as from step 1, in order to obtain further PLS components. The predicted values of y for t components are given by the sum of y_t over these t components.

Index